IMAGES
of America

LOGGING IN GRAYS HARBOR

On the Cover: The unimaginably large coastal Washington State trees challenged early Grays Harbor hand loggers in their efforts to "bring in the wood." Around 1900, seven pioneer hand loggers tackle a giant using axes and saws powered by their own bodies. Notches were cut into trees, into which were inserted 12 springboards, overlaid with 9 others, forming a chopping platform. This method raised the cut above the wide "skirt" of the trunk, with its crooked grain. The horizontal, flat undercut was made by using a two-man, long, falling saw. This was followed by using a long-handled falling axe to chop out a wedge-shaped undercut area in the direction the loggers wished to fall the tree. Opposite, and three inches above the undercut, the hand-powered falling saw began the hard work of cutting through the diameter of the giant trees. It was a hard day's work indeed. (Aberdeen Museum of History.)

IMAGES
of America

LOGGING IN GRAYS HARBOR

Brian Woodwick and Gene Woodwick

ARCADIA
PUBLISHING

ISBN 978-1-4671-3189-6

Published by Arcadia Publishing
Charleston, South Carolina

Printed in the United States of America

Library of Congress Control Number: 2013955120

For all general information, please contact Arcadia Publishing:
Telephone 843-853-2070
Fax 843-853-0044
E-mail sales@arcadiapublishing.com
For customer service and orders:
Toll-Free 1-888-313-2665

Visit us on the Internet at www.arcadiapublishing.com

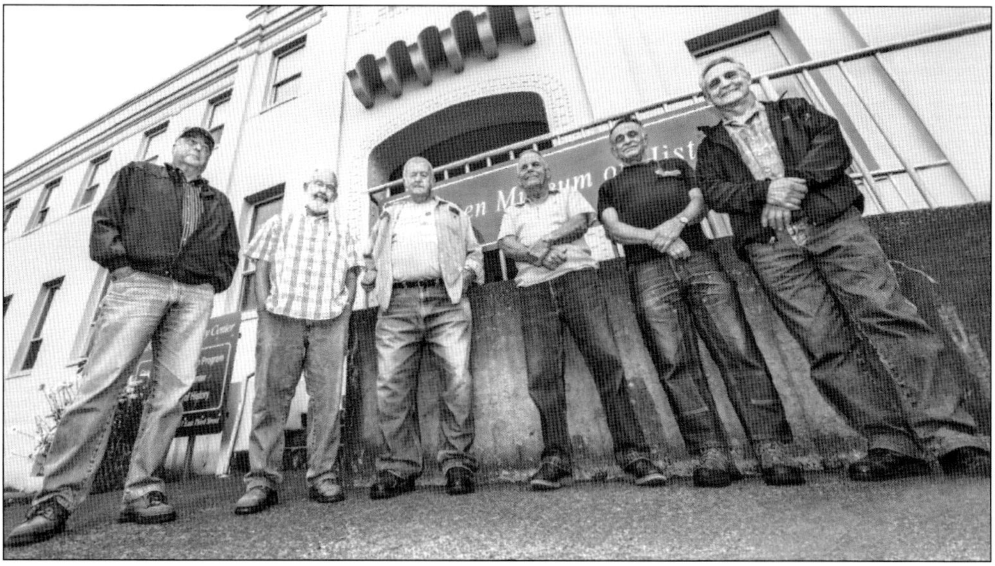

This book is dedicated to the skilled men of the Grays Harbor logging community, who exemplify the heritage of hard workers committed to their workplaces; good citizens who have contributed greatly to the community's well being; caring and loving family men and men of honorable character. It is working loggers like them who helped create a proud heritage for Grays Harbor throughout more than a century of logging to supply the material needs of people around the world. They stand tall. God bless them all.

Contents

ACKNOWLEDGMENTS

Those dedicated to the preservation of Grays Harbor heritage and contributing images are the following: director Dann Sears, Aberdeen Museum of History; director John Larson and Scott Lucas of the Polson Museum; Joseph Shay, archivist, and Lisa Benn, Port of Grays Harbor; Kelly Calhoun and Lee Merrick, Museum of the North Beach; Larry Workman, photographer, Quinault Department of Natural Resources; Peter Darrrah, director, Willapa Seaport Museum; and Jarrod Todd, Robert Reime, Paul Mackenzie, and John Henson of Logger's Playday 2013. Those who have made the book possible are Jolene Rabey, Rabey Trucking, Inc., Tom Quigg, Byron Eager, Ralph Larson, Roy Vataja, Esther Kinney, Jeffery Dukes, and Bernie Paul. Many photographs in the book are from the collection of Gene Woodwick (Coastal Heritage Services) and Brian Woodwick (Brian Woodwick Photography), who served as documentary photographer and performed photograph restoration. Photographs even with incomplete identification have been included for the enjoyment of the reader. Sadly, hundreds more photographs had to be eliminated due to space limitations.

The following image sources appear as abbreviations in the courtesy lines: Aberdeen Museum of History (AMH), Coastal Heritage Services Gene Woodwick Collection (CHSGWC), Dale Hill Collection (DHC), Museum of the North Beach (MNB), Pat Rabey Trucking, Inc. (PRT), Polson Museum (PM), Quinault Department of Natural Resources (QDNR), and Willapa Seaport Museum (WSM).

INTRODUCTION

Grays Harbor, with its numerous tidewater streams running through amazingly dense forests, was, and is, timber country. The county grows and harvests trees just as Kansas does wheat. The 90-square-acre harbor is situated like a huge mouth smack in the middle of the Washington State coastline—the perfect place for a port situated three days closer than any other state port to its traditional US, Asian, and other foreign markets.

Homesteaders switched from farming to forests in 1880, hitching up the family horse to hand-cut logs for market. By 1899, horse loggers had been replaced by oxen skinners, whose profanity was without measure in the known world. Early loggers had caught the attention of one Capt. Asa Simpson of San Francisco, who sent his mill manager, George Emerson, to the developing town of Hoquiam in the Washington Territory in 1882. Soon, Emerson bought up an entire sawmill and loaded it on the company's schooner, *Orient*, to mill the massive spruce, red cedar, and hemlock trees. By 1899, ox loggers, benefiting from easily accessible river and harbor frontage, had an underestimated output that totaled 185 million feet of logs, from 40 camps operating on five rivers (Hoquiam, Humptulips, Wishkah, Chehalis, and Nueskah). Their output provided the impetus for a milling industry that produced lumber, shingles, boxes, spar poles, furniture doors, and more. The manufacturing base later included veneer, pulp, plywood, specialty papers, and unique items, such as airplane parts for Charles Lindbergh's *Spirit of St. Louis* and the wood for the piano of Paul McCartney of The Beatles.

Animals were replaced by donkey engines, which took wire rope up to 3,000 feet into the timber to haul logs to a vast network of skid roads, rails, splash dams, roll dams, and river waters, by which the logs reached the mills. Steam had come to the woods. By the beginning of the 1900s, 52 steam donkeys valued at $156,000 each were at work in camps averaging 75 men. The timber industry remained the economic driver for the county throughout the 20th century.

The first Grays Harbor loggers cleared massive firs, cedars, and spruce off farmlands; they were quickly followed by men from Nova Scotia, Canada, the forests of New England, and the Great Lakes region who heard about the new Washington Territory's coastal areas, with their vast timber tracts. They trekked westward on foot, horseback, and sailing ships to make their fortunes in the green gold of places like the Promised Land of the Humptulips Valley. There, trees grew so thick that, upon being cut down, they often did not land on the ground. Today, five generations later, the area still produces trees for the logging industry.

From the beginning, camps sprang up like mushrooms to house and feed workers and to supply them with oxen, horses, and machines to bring daylight to the swamps. Among the largest camps were Brooklyn Saginaw Camp; Greenwood, owned by Bill Boeing; Charley Clemons's camp, which became the first recognized tree farm in the Northwest; Grisdale, of Simpson Timber Company fame (last of the family logging camps); and Schafer Brothers, which built a family-owned empire. Polson had its famous family Railroad Camp. Camp 14, Crane Creek, and other camps housed single men only. At some sites, men went home to their families on the weekends. And there was the infamous Grays Harbor Lumber & Mill Company, referred to as the "Penitentiary of the West" for its ill treatment of workers.

Camp life built heritage. It was the supreme college of logging skills. It cemented long-lasting friendships, marriages, and family life. It was not an isolated way of life, but one that took in loggers from all over the Olympic Peninsula on a regular basis, as specific logging skills or equipment were needed on a site. It took tough men with the ability to anticipate danger, to see with "eyes in the back of their heads." The loggers worked 10 or 12 hours on hoot owl (during the night to eliminate the danger of fire) or to know it "better come or bleed," be it big log or heavy equipment because these men had endurance, strength of body and mind, and often used words that could curl the bark off a cedar tree.

But, good or bad, the men came to work in the woods by the thousands, and some never went back home. From 1920 to 1929 alone, over 1,700 men lost their lives in this dangerous occupation; in Washington State between 1880 and 1999, a total of 8,000 loggers were killed. But for any kid who grew up in a logging camp or lived with a "Pa" or "Gran'pa" who worked in the camps, there was only one goal in life—to become a logger.

Robert F. Lytle kicked off the 20th century as the most extensive logging firm on the harbor. Mack Logging Company was the first camp to log entirely by railroad. The largest timberland owner was R.D. Merrill (later Merrill & Ring). Hoquiam was the headquarters for all, including the Larkin Brothers, Ellison & Soljeth, Davis & Dineen, and the camp of Reuther & Snow. The Polson Company, owned by Alex and Robert Polson, absorbed all the pioneer outfits, plus more, to become the largest logging company in the world. In turn, it sold out to Rayonier, Inc., in 1948, which had purchased the Bloedel-Donovan operations in 1945. Clemons Logging became a subsidiary of the Weyerhaeuser Company (Weyco) in 1919 and conveyed its lands to them in 1936. Weyco had entered the area by buying out the holdings of the Grays Harbor Commercial Co. at Cosmopolis, converting it to a woods and pulps mill complex. Those changes led to the corporatization of Grays Harbor with outside, not local, interests. Asa Simpson's brought his company from San Francisco to Hoquiam, then expanded to southwest Washington and Oregon. In the eastern part of the county, another Simpson, Sol, set up Shelton headquarters, where it holds sway today throughout the Northwest under the Green Crow label and is managed by Sol's descendants.

Grays Harbor, always a hotbed of unionization, has a long and varied history of ordinary loggers and mill workers fighting to improve daily living conditions, safer working conditions, and better wages. All the big names of labor organizers were involved. The "Wobblies" of the International Workers of the World (IWW) were in evidence, especially in the Finnish community of workers. The red-haired, petite Elizabeth Gurley Flynn mounted soapboxes on the corners of Aberdeen streets to urge the wood workers to unite. The formation of the American Federation of Labor (AFL) and the Congress of Industrial Organizations (CIO) were strife-ridden processes. At the outbreak of World War I, the Wobblies pulled off the biggest work strike ever known and shut down the woods. The need for spruce during the war led the government in 1917 to bring in the Loyal Legion of Loggers and Lumbermen (Four-L) under the leadership of Col. Bryce Disque, in charge of the newly formed Military Spruce Division. His officers refused to put their men into the horrible work and living conditions that existed in the camps. Disque ordered the drive for eight-hours days, minimum wage, clean living conditions, better food, and safer working conditions, thus fulfilling all of the IWW goals. Regardless of the progress, local loggers referred to the 4Ls as the "Lazy Loggers and Loafing Lumbermen."

Following the war, lumber prices plummeted and jobs were lost. The industry was saved by the 1921 earthquake and tsunami that devastated Japan, leading to a need for vast amounts of lumber. The federal decision in the 1980s on the spotted owl spelled the end of logging as Grays Harbor had known it, leaving four and five generations of families to move and find work elsewhere. But, oddly enough, it was again an earthquake and tsunami that revived the near-dead industry in 2011.

After World War II, logging provided hundreds of families with a lifestyle on a par with the city's corporate families. In addition, the heyday of logging camps provided richness in pride of work ethic, lasting friendships, and a tight-knit community of housing, homes, and bosses who shared the life of the workers. It was the era when steam left the woods and lumber products took to the roads in trucks. Finally, the corporations phased out the family camps; the outlying camps closed, and folks moved into the core areas of small towns and cities.

Still today, most Grays Harbor folks remain proud of being a part of the heritage of logging and its associated industries. And the words "Grays Harbor" are still synonymous with loggers and logging.

Today, all logging begins by cruising timberlands to estimate the value of merchantable timber by species and volume. Cruising sets legal boundaries, is money in the bank for a timber contract loan, and differentiates government lands from corporate lands, private lands, and gyppo (private, non-corporation) logger lands.

Without cruising, there would be no viable loggers, no timberlands, and no money for county, state, or federal tax monies for anticipated budgets. It all began with an Irish immigrant, Jeremiah "Jerry" Ambrose McGillicuddy, born in 1852. He was logging on Smith River when Asa Simpson sent George Emerson to purchase a mill site in Hoquiam. Jim Cowan was ordered to log the Hoquiam River with oxen. McGillicuddy built a homestead at the tidewater head of the river and married Cowan's daughter Agnes, which positioned him to begin locating homesteads with Alex Polson in 1885, when Polson began logging with oxen; the two, with Dan Dineen, built the area's first splash dam in the harbor, near New London.

McGillicuddy cruised the entire Chehalis County (now Grays Harbor) before 1912. Between 1909 and 1913, as chief cruiser for Grays Harbor County, he managed the first cruise in the state for a more equitable evaluation of forestlands. During World War II, he scouted out the best grade of spruce in Grays Harbor and Pacific Counties, free of charge, as his contribution to the war effort. McGillicuddy's son Blaine returned from that war and began his lifelong career as a cruiser. Today, four generations of McGillicuddy offspring are Grays Harbor timber folks, connected to timber history through various enterprises, from shipping, to real estate, to various types of construction.

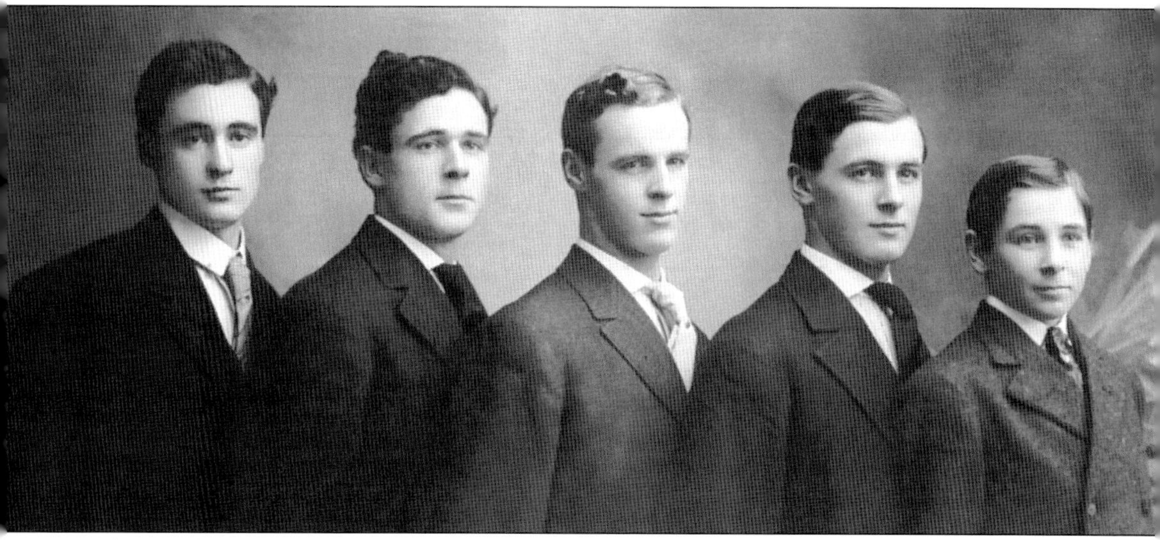

The fourth-generation progeny of Jerry McGillicuddy are still heavily invested in the future of timber and Grays Harbor. His children are, from left to right, Agnes B., Jeremiah Ambrose Jr., J.E.C.F., F. Daniel, Eudene M., and Blaine H. (Tom Quigg.)

One

First Loggers

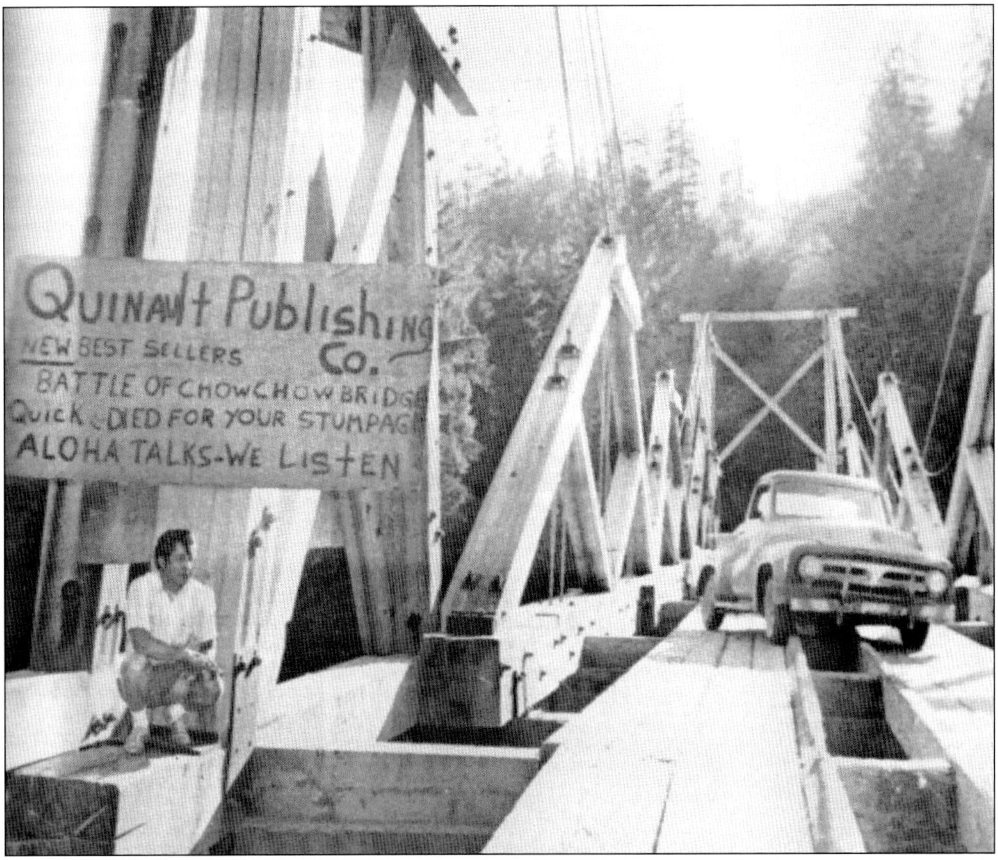

Joe DelaCruz, Quinault tribal chairman, blockaded the Chow Chow Bridge over the Quinault River on September 13, 1971, to protest the Bureau of Indian Affairs' (BIA) allowance of a 30-percent reduction in stumpage prices from the Aloha/Evans and ITT Rayonier, Inc., to the Quinault Nation. Following the blockade and four lawsuits against the BIA, the Quinault's right to manage its own natural resources was established. (QDNR.)

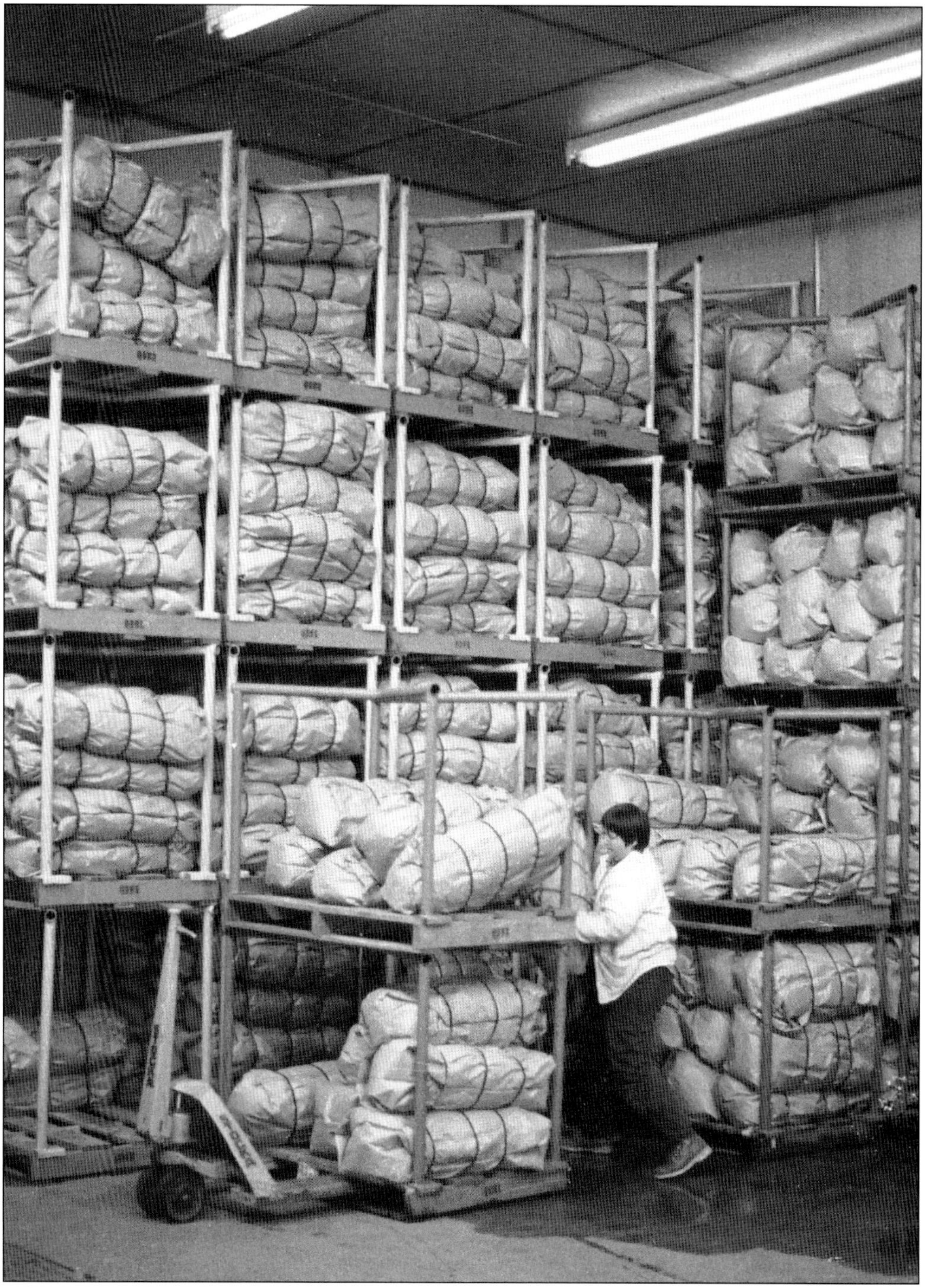

After 20 years of non-Indian, commercial logging under BIA management, from 1960 to 1971, less than one percent of reservation lands had been replanted, greatly affecting the Quinault Nation's fisheries. In 1986, Cheryl Bumgardner checked seedlings for the spring plant. In the early 1920s, the tribe planted over 1,000 acres of seedlings in an "experimental forest"—the first Grays Harbor tree farm. (QDNR.)

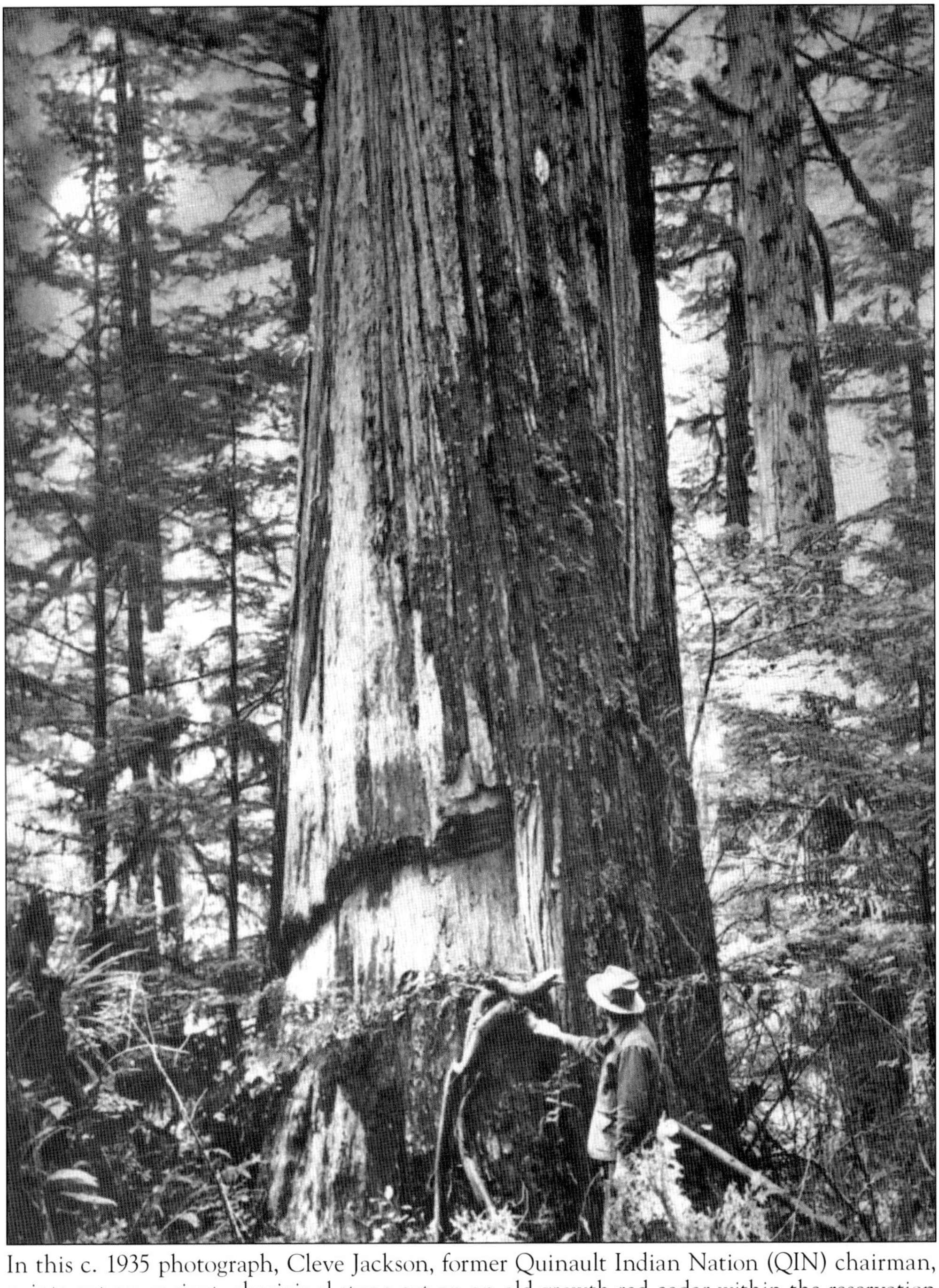

In this c. 1935 photograph, Cleve Jackson, former Quinault Indian Nation (QIN) chairman, points out an ancient, aboriginal stone cut on an old-growth red cedar within the reservation boundaries. Following the Bodlt decision of 1974, confirming the right of Indians as comanagers of resources, the Quinault Nation developed its own forestry department. (QDNR.)

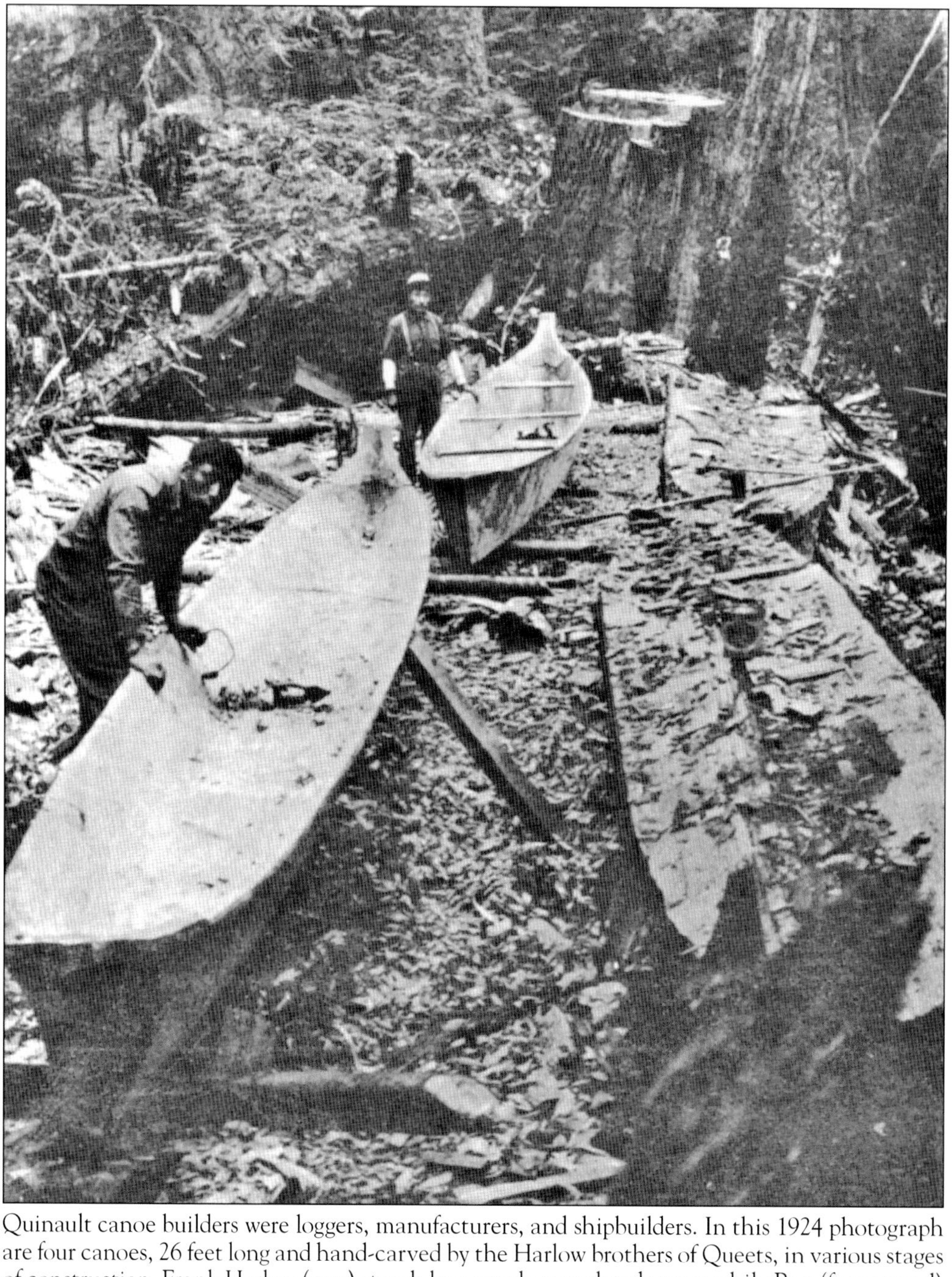

Quinault canoe builders were loggers, manufacturers, and shipbuilders. In this 1924 photograph are four canoes, 26 feet long and hand-carved by the Harlow brothers of Queets, in various stages of construction. Frank Harlow (rear) stands by a nearly completed canoe while Ben (foreground) shapes the interior of another; two rough-cut canoes await completion. The four canoes were made from one cedar tree, logged during the construction of the future route of US 101, from Grays Harbor to Forks. (CHSGWC.)

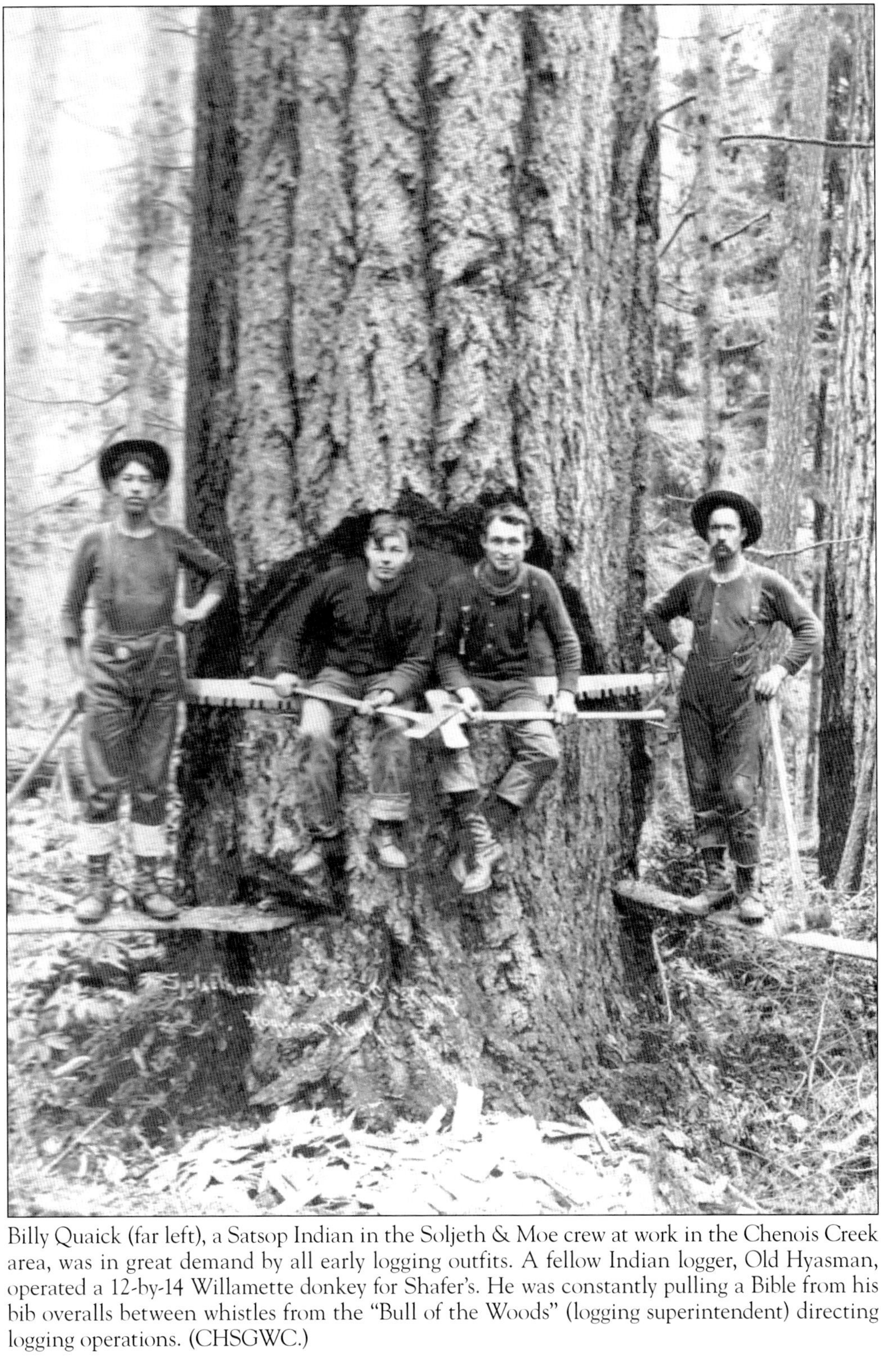

Billy Quaick (far left), a Satsop Indian in the Soljeth & Moe crew at work in the Chenois Creek area, was in great demand by all early logging outfits. A fellow Indian logger, Old Hyasman, operated a 12-by-14 Willamette donkey for Shafer's. He was constantly pulling a Bible from his bib overalls between whistles from the "Bull of the Woods" (logging superintendent) directing logging operations. (CHSGWC.)

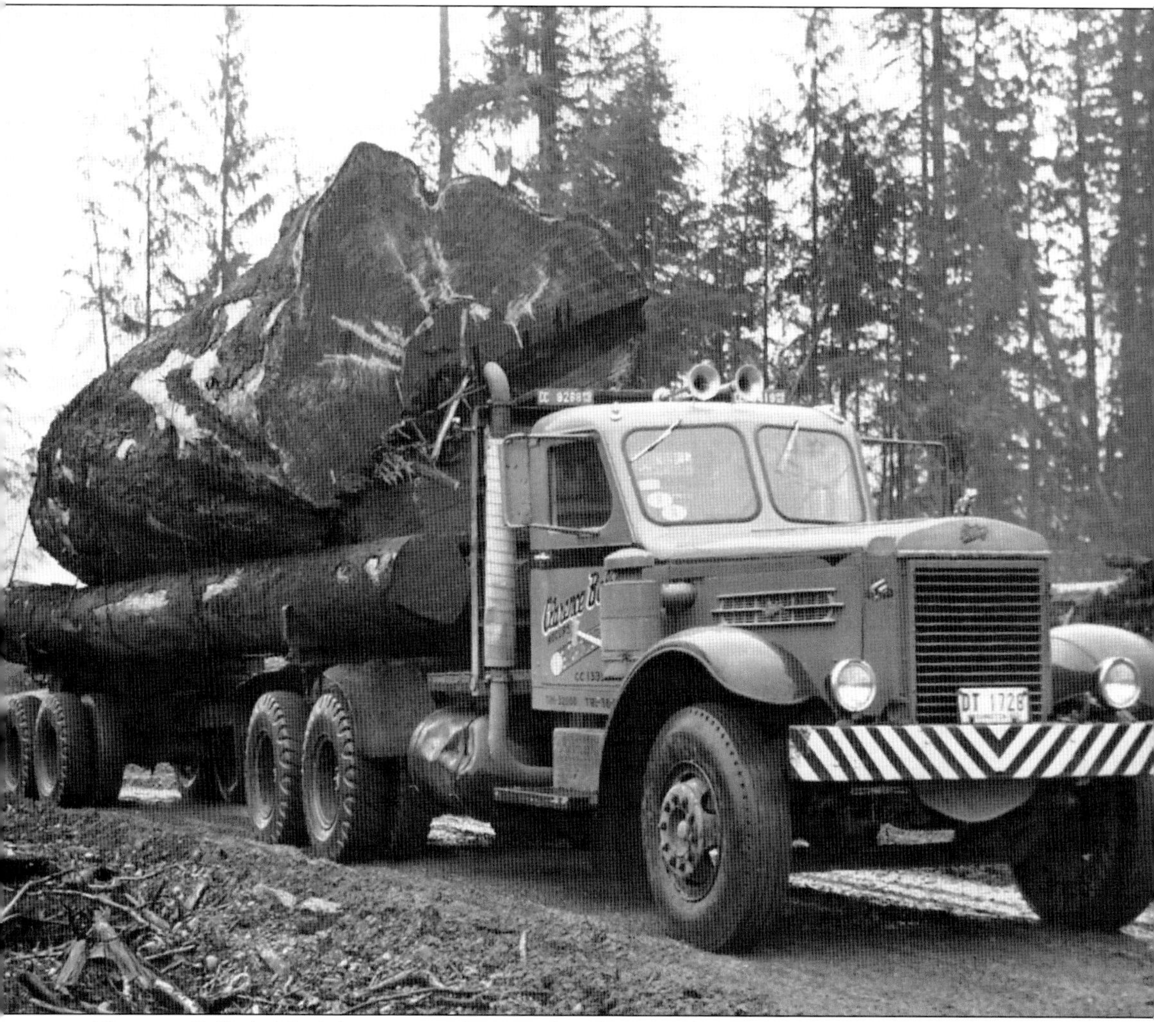

Clarence Baller's Mack logging truck was a familiar North Beach sight. He was a truck driver for a variety of Bureau of Indian Affairs contract loggers, such as Aloha and Mayr Brothers on the Quinault Reservation. Timber sale rights passed from Aloha to M.R. Smith to Harbor Plywood and then to Mayr in 1979. Now, all of the logging companies and the town of Aloha are gone. (NBMH.)

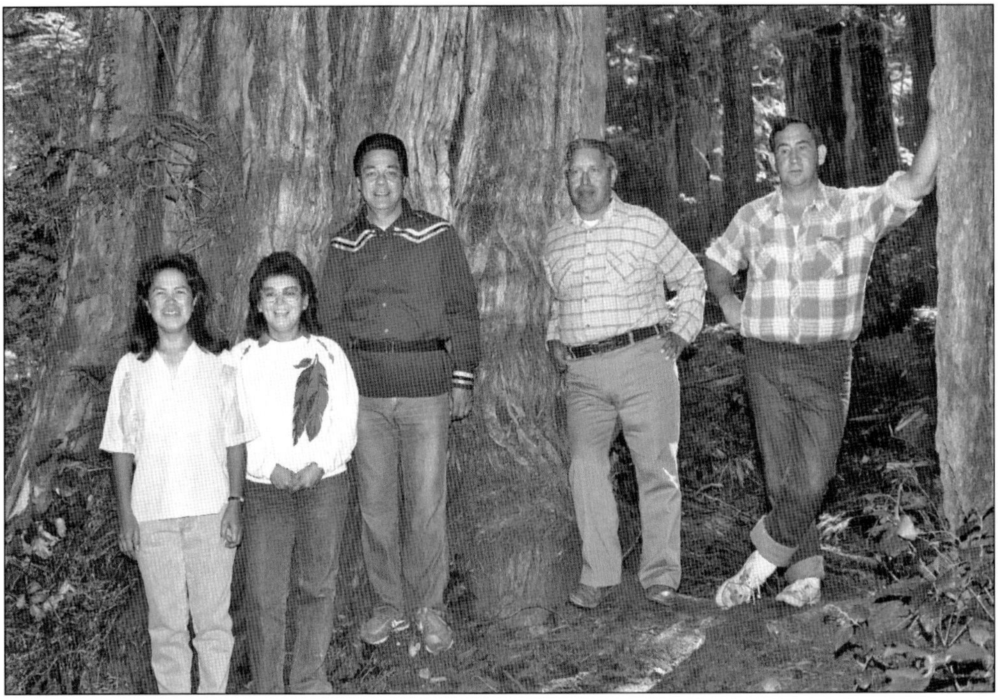

In 1987, Quinault tribal officials pose before one of the last groves of red cedar on the reservation. Shown here are, from left to right, Pauline Capoeman, secretary and director of the Quinault Department of Natural Resources; Pearl Capoeman-Baller, QIN vice chair and administration director; Joe DelaCruz, QIN president; Phil Martin, first councilman and Cook Creek Hatchery director; and Warren Shale, QIN treasurer. (QDNR.)

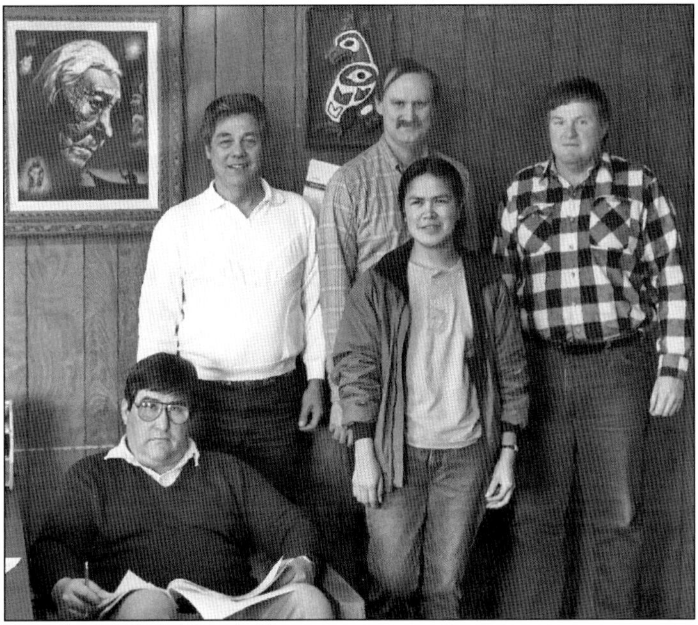

The Iran-Contra hearings forced negotiations with the tribe for 9,985 acres of land that Gen. Richard Secord had attempted to purchase with profits from the Iranian arms deal. The Quinault Land & Timber Enterprise (QLTE) was developed to operate business ventures related to forest resources. Shown here are, from left to right, director Guy McMinds (seated), president Joe DelaCruz, general manager Alan Steege, QDNR director Pauline Capoeman, and manager Larry Black. (QDNR.)

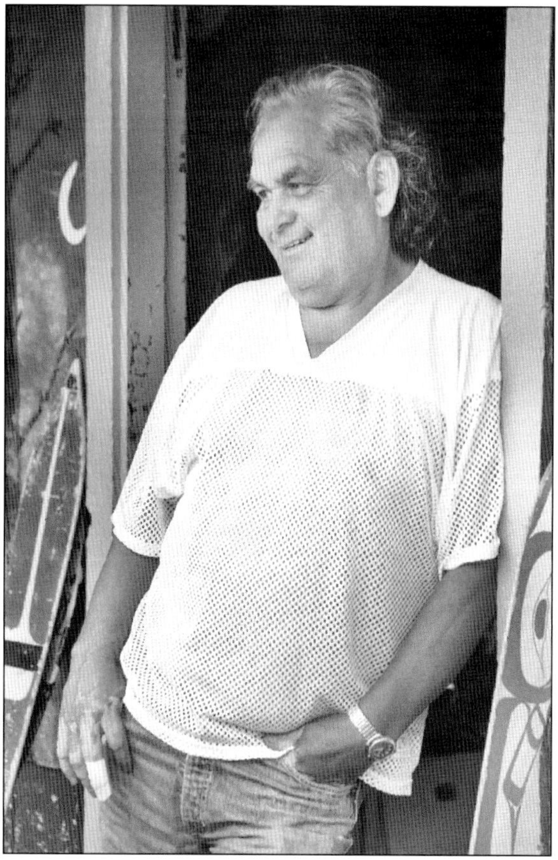

Justine James Sr. set an unbeatable record in 1963 by cutting 70 squares in a six-hour shift at the L.J. Esses Shingle Mill at Lake Quinault, using the skills he learned from 70-year-old Ray Sharp. Later, the mill became Esses Logging Company at Amanda Park. Beginning at the age of 21, James participated in 43 years of involvement in tribal affairs. (Brian Woodwick.)

Justine James Jr. began his career in natural resources working in fire-suppression Indian "Hot Shot" crews. He saw action during the 1991 Yellowstone Fire and later worked in the Department of Natural Resources as a Timber, Fish & Wildlife compliance officer. He is an author and Quinault cultural analyst. (Brian Woodwick.)

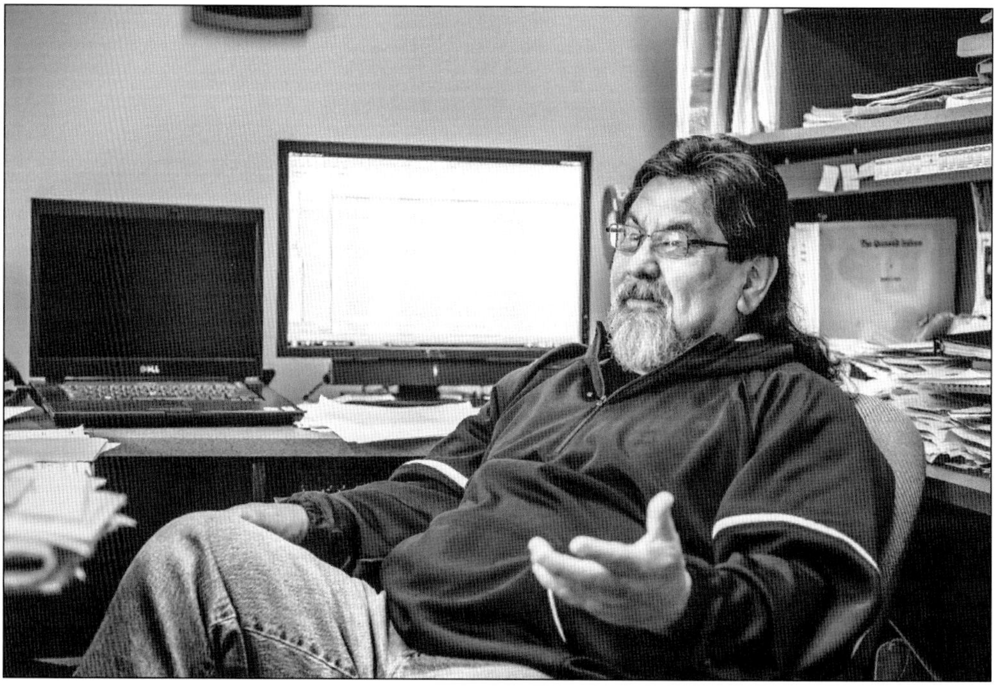

Two

Homestead Horses and Bullwhackers

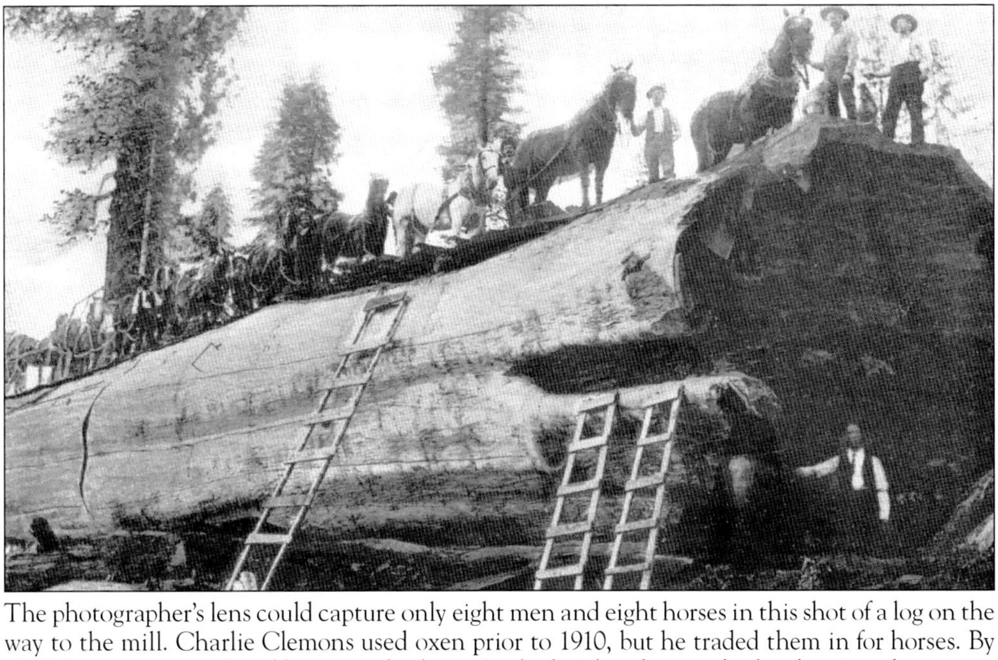

The photographer's lens could capture only eight men and eight horses in this shot of a log on the way to the mill. Charlie Clemons used oxen prior to 1910, but he traded them in for horses. By 1900, horses were replaced by steam donkeys. At the height of steam donkey logging, the county boasted 900 of the steaming, snorting engines. (GHSGWC.)

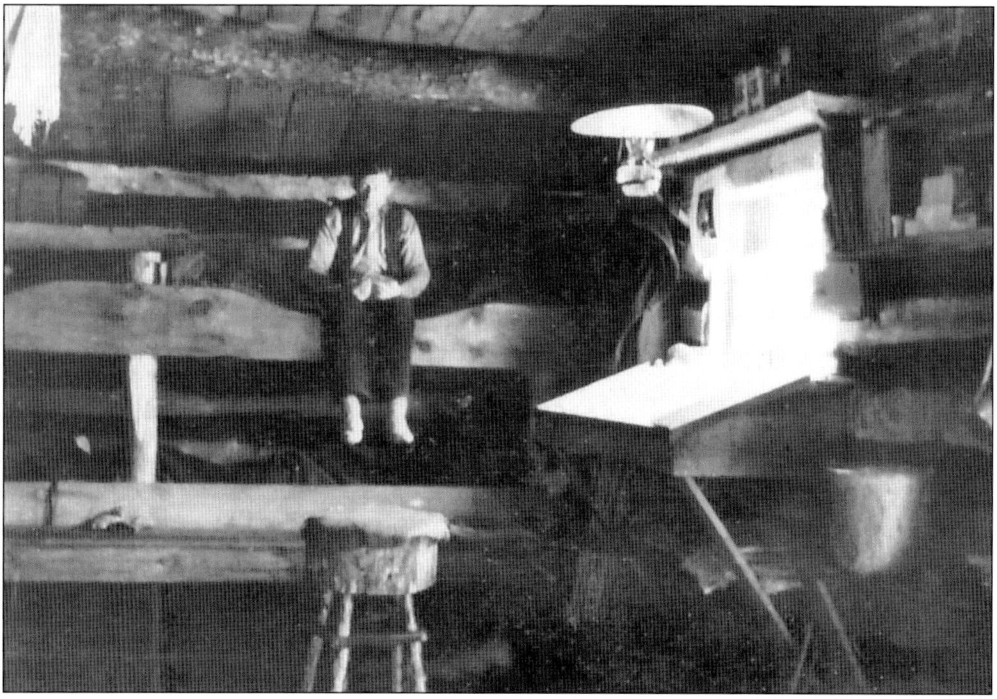

This early photograph postcard, sent home to mom, carries the following message: "This is our home sweet home for the winter. My bunk is the lower one. The boss starts every day drunk." Considering that the working day was 10 hours long, it was no wonder the logger was disgruntled. (CHSGWC.)

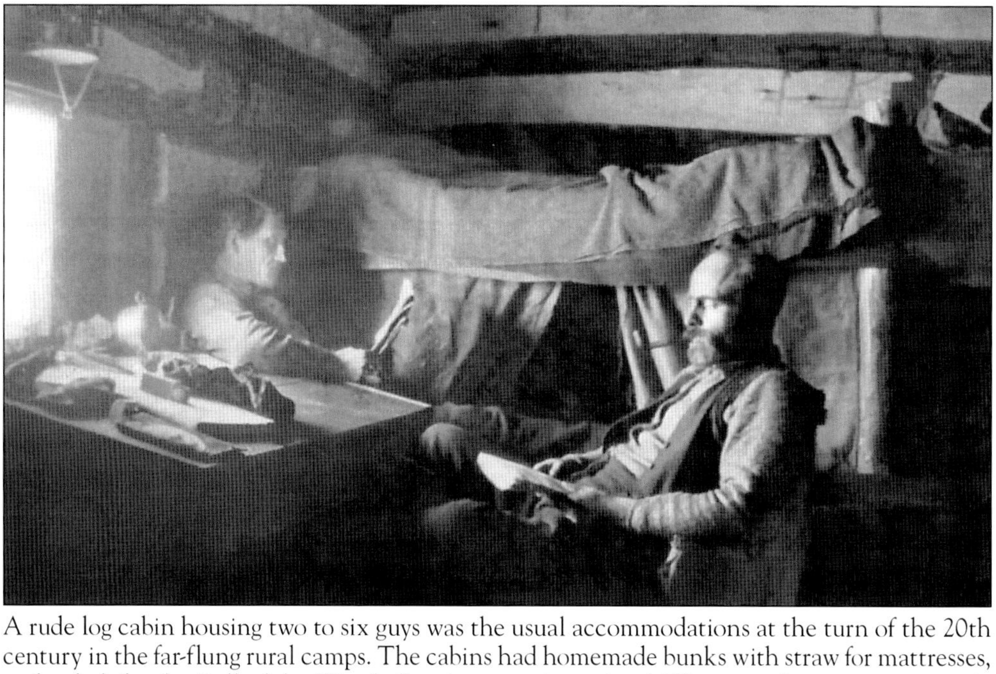

A rude log cabin housing two to six guys was the usual accommodations at the turn of the 20th century in the far-flung rural camps. The cabins had homemade bunks with straw for mattresses, and a desk for the Bull of the Woods (logging superintendent). These rural camps were mostly owned by individual homesteaders. (CHSGWC.)

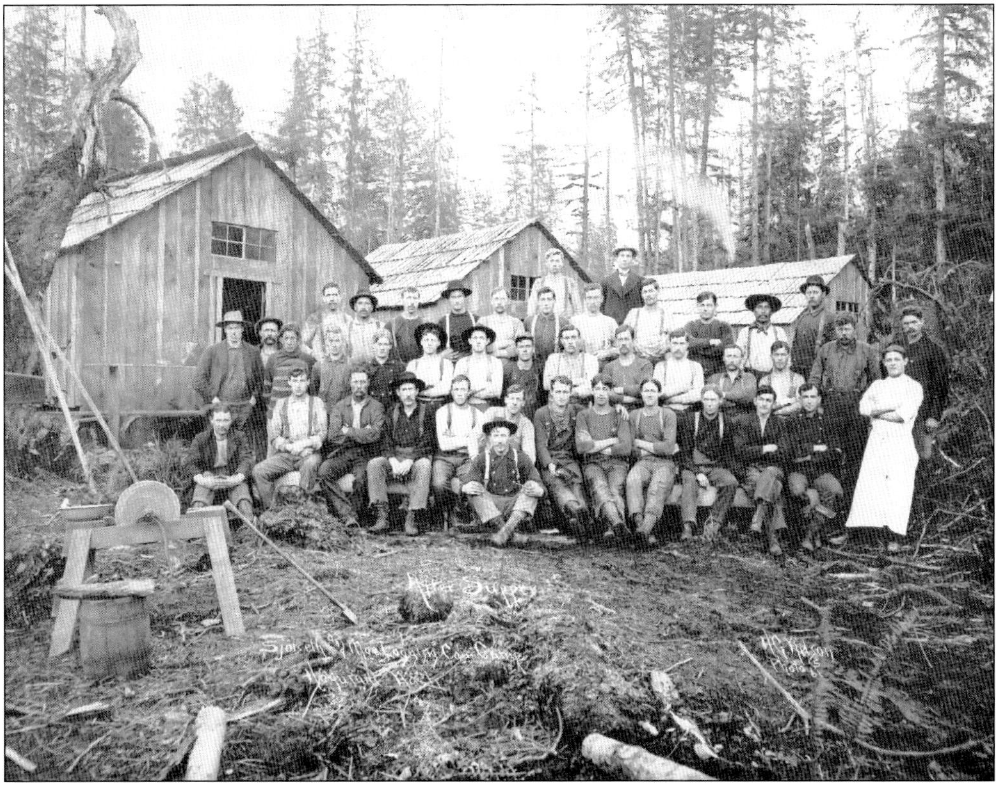

The Gillis camp crew is pretty hungry, waiting for supper on June 6, 1901. They had logged a fir tree that produced nine saw logs, of which seven were 24 feet long; two others were 28 feet long. The butt of one of the logs measured 60 inches in diameter at the small end. (CHSGWC.)

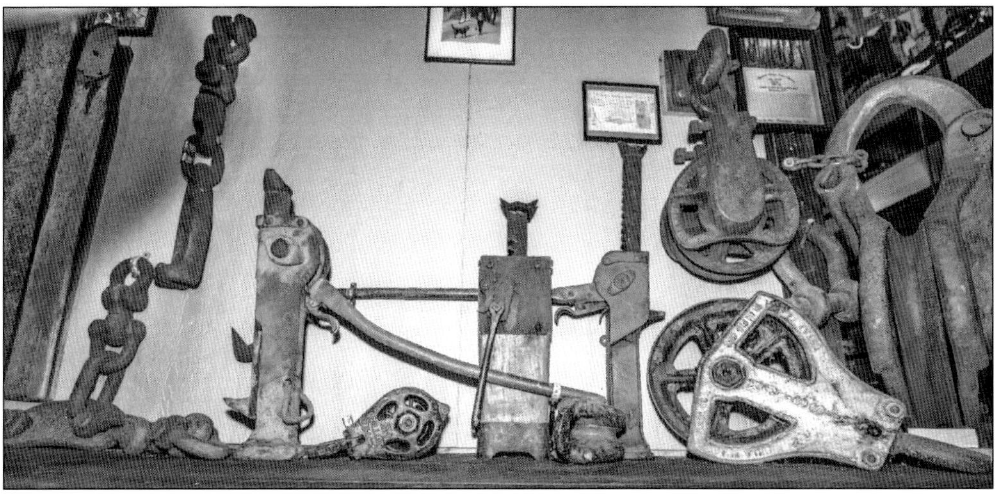

Plying crankjacks, like these on display at the Willapa Seaport Museum, was a dangerous job, as it forced men to stand close to piling where they could easily slip, resulting in injury or death. Shown here are the Willapa Iron Works Gilcrest jacks and high lead, self-oiling Hercules blocks. From left to right are snaffle for hanging clocks, Hercules jack, unidentified brand jack, 1910 crankjack, Gilchrist crankjack, 1910 block (upper), Tulsa block (lower), and a Gilcrist self-oiling bull block. (Brian Woodwick.)

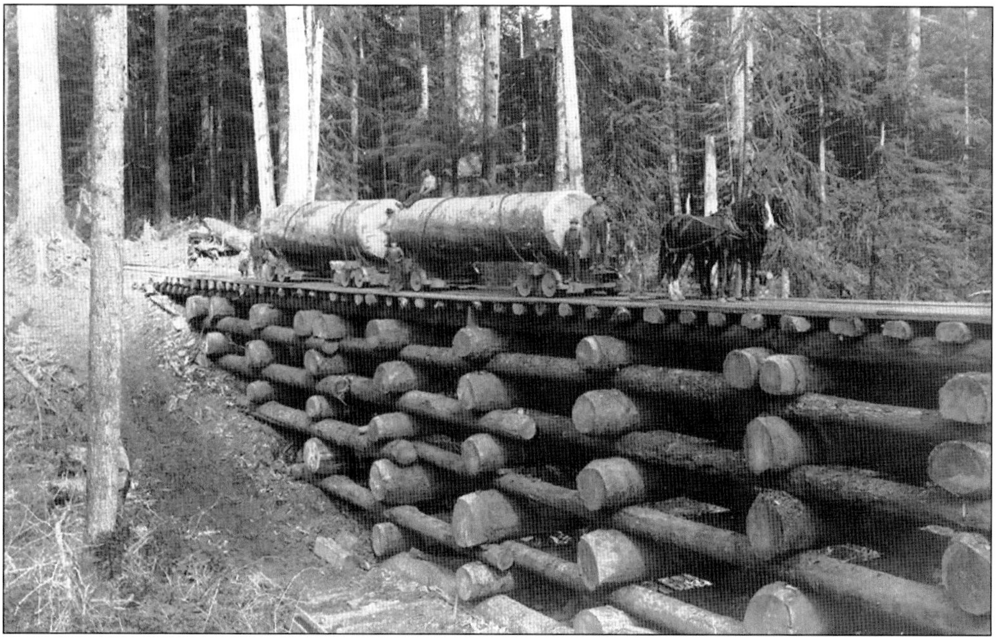

Typical farm workhorses, not big Clydesdales, pulled sniped logs on truks (a vehicle with two sets of wheels without a middle connection). The truks rode narrow log "rails" over a cribbed trestle. The average ox team consisted of four teams of yokes. Heavy loads required seven to nine teams. (CHSGWC.)

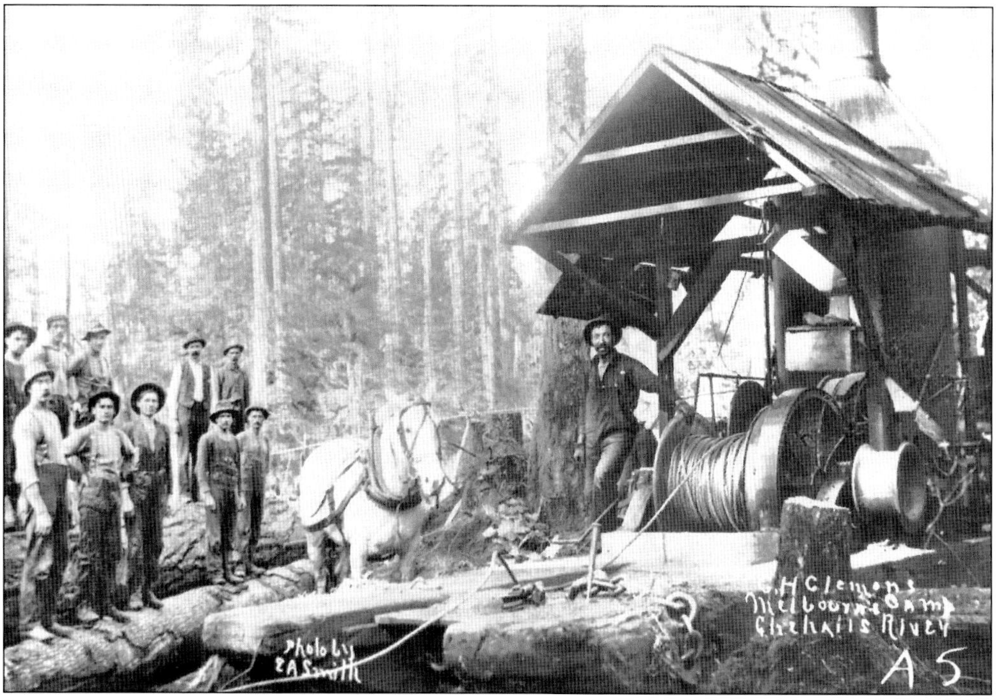

This C.H. Clemons crew, located at Melbourne on the Chehalis River, used a line horse to yard logs connected to the spool drum donkey. The Melbourne camp was the largest of the Clemons camps at that time. (CHSGWC.)

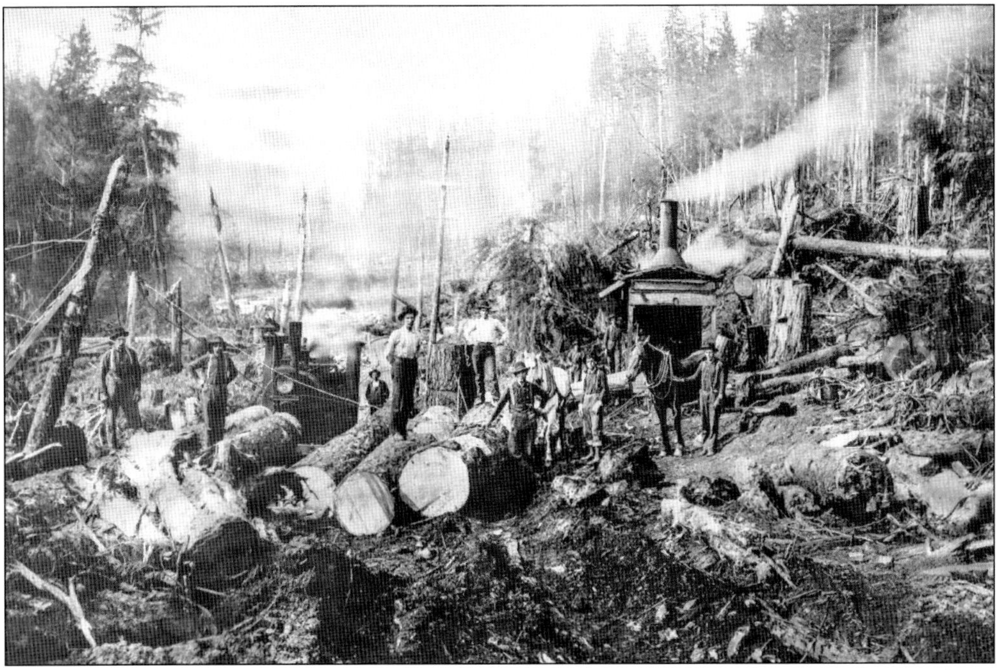

The character of some logging shows was such that a few camps used a haulback or trip line drum on their donkeys, with the line carried entirely by horses. The early day line was made of manila rope, later replaced by wire of iron, and then replaced by steel strands. (CHSGWC.)

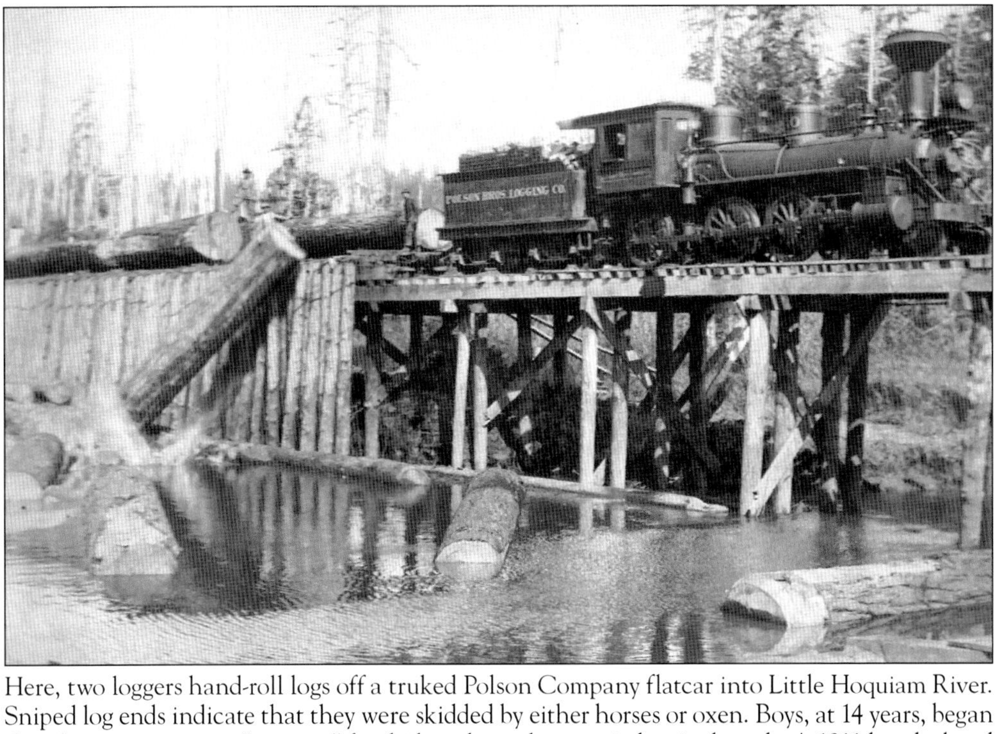

Here, two loggers hand-roll logs off a truked Polson Company flatcar into Little Hoquiam River. Sniped log ends indicate that they were skidded by either horses or oxen. Boys, at 14 years, began their logging careers as "greasers" for skid roads, or they carried springboards. A 1911 law declared that "none under the age of 18 years" was to be hired as firemen on donkeys. (CHSGWC.)

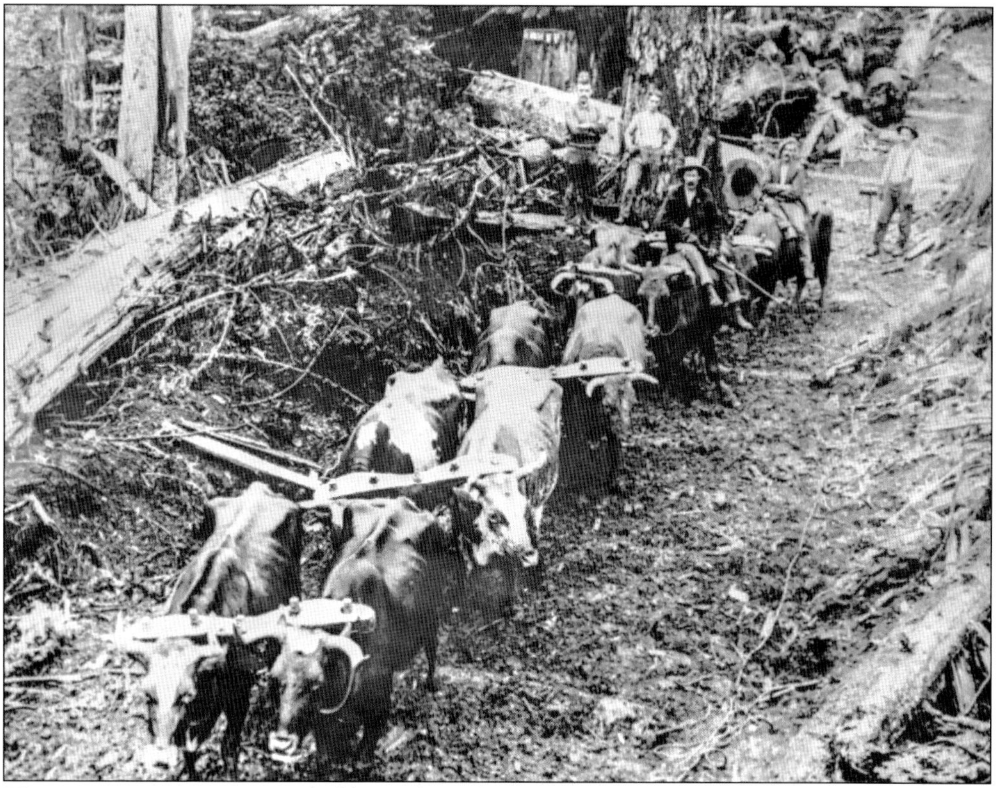

The average ox team consisted of four yokes. The massive Grays Harbor trees often limited loads to just one log section, requiring seven to nine teams and poked by "go devils" (goads) and plenty of cusswords. Oxen also yarded logs using 12-to-15-foot chains attached to the yokes. The chains had a "grab" hook and/or a slip hook to connect to the logs. (AMH.)

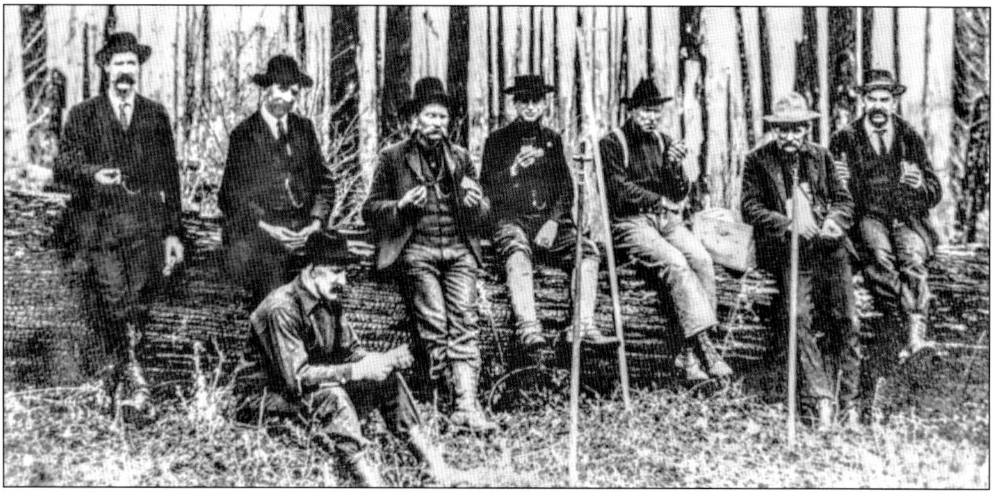

The famed early pioneer loggers of Grays Harbor included, from left to right, C.H. Clemons, William Maloney, George Arland, (seated in front), Joe Bernard, E.K. Bishop, Dan McCroskey (Maloney's brother-in-law), Ben Kesterson, and George Ninemire. Clemons eventually became one of the county's largest timber outfits. Bishop left behind a legacy of philanthropy, and Ninemire became a well-known logging camp purveyor. (AMH.)

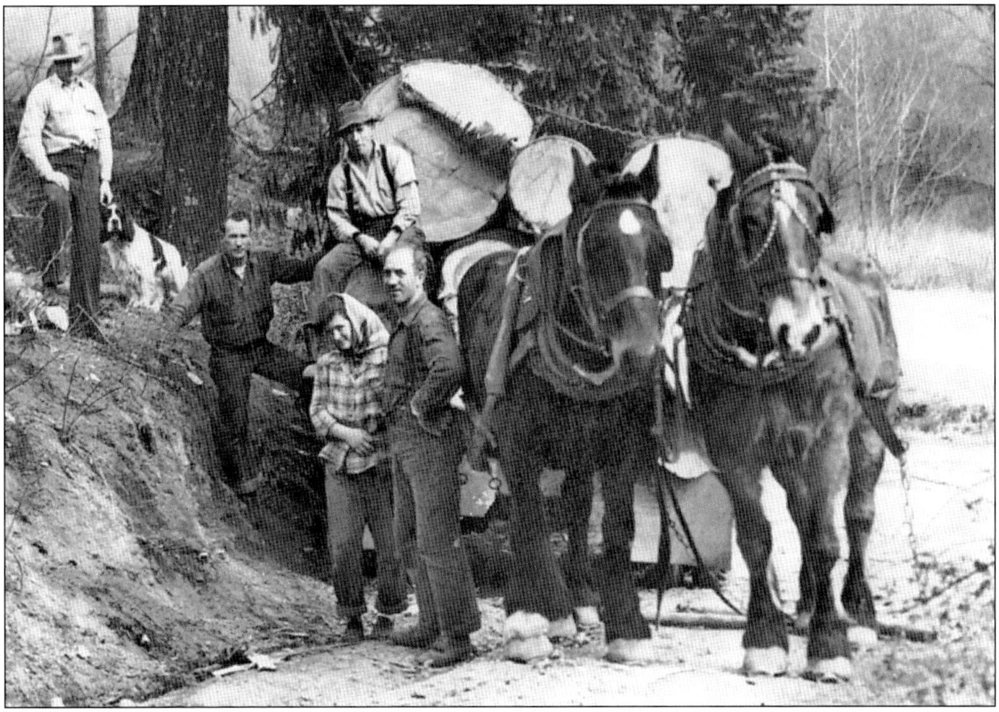

Following World War II, horses were in use by small gyppos. Here, Mr. and Mrs. Red Jinley (in front) pose with their crew. Steve Withers is on the log, an unidentified logger is on the bank, and a Mr. Morganthaler keeps a firm hand on the camp dog. The mid-1960s saw a revival of horse logging, following increased environmental concerns. (CHSGWC.)

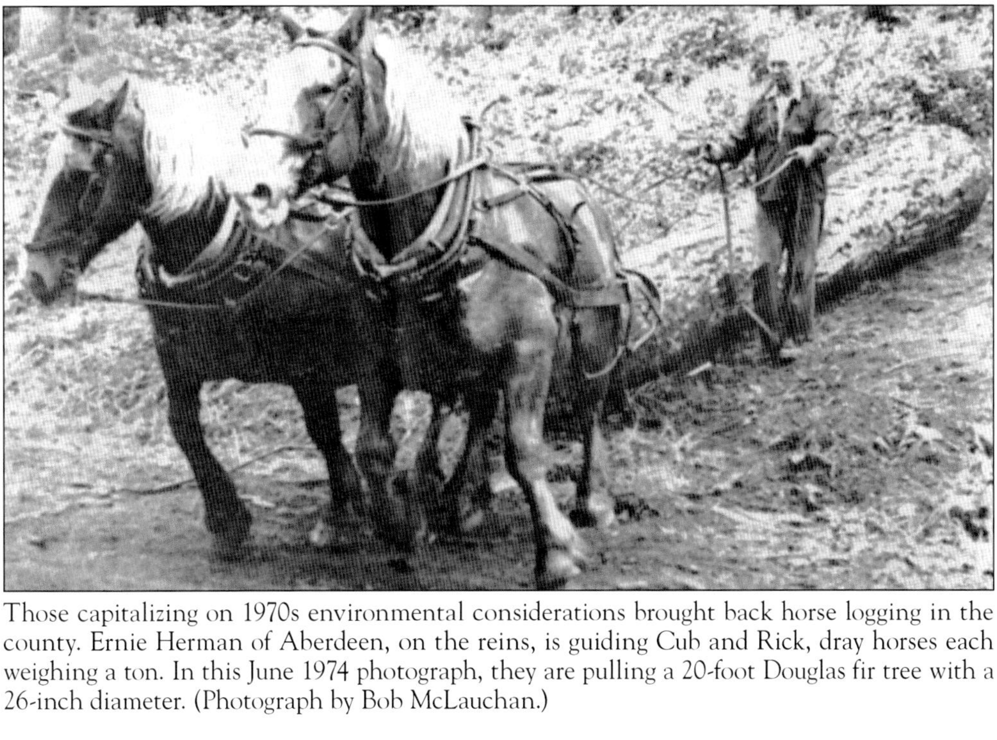

Those capitalizing on 1970s environmental considerations brought back horse logging in the county. Ernie Herman of Aberdeen, on the reins, is guiding Cub and Rick, dray horses each weighing a ton. In this June 1974 photograph, they are pulling a 20-foot Douglas fir tree with a 26-inch diameter. (Photograph by Bob McLauchan.)

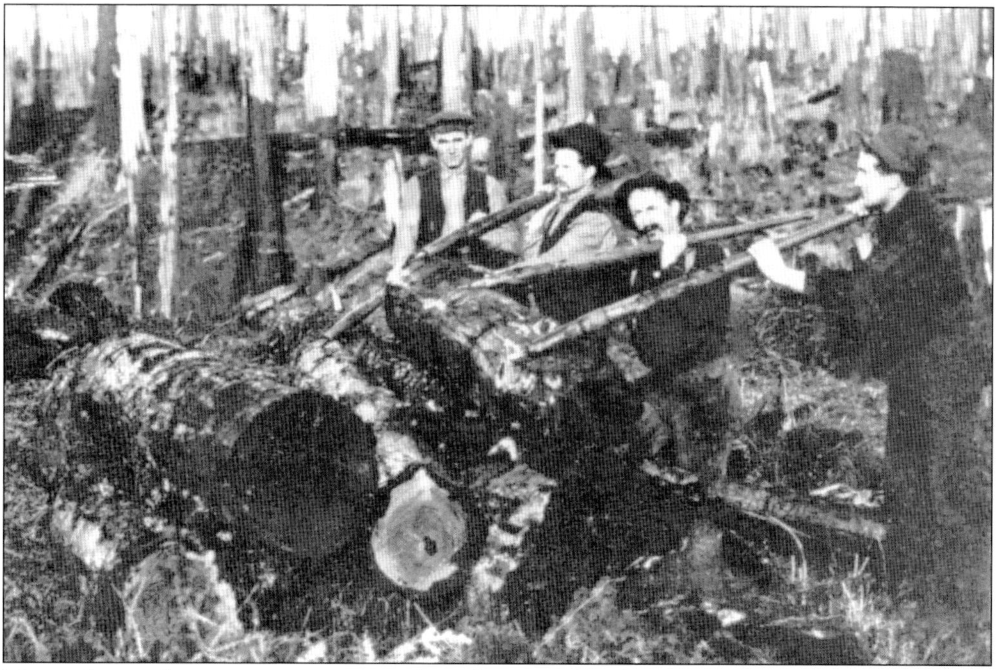

Moving logs with peaveys was no easy task, as shown in this photograph. The Willapa Iron Works production of crank jacks was a great improvement for moving logs from the woods to the rivers for transporting to the area's mills. The company's Gilchrist jack, wedges, and blocks became logging staples all along Washington's coastal logging areas. (CHSGWC.)

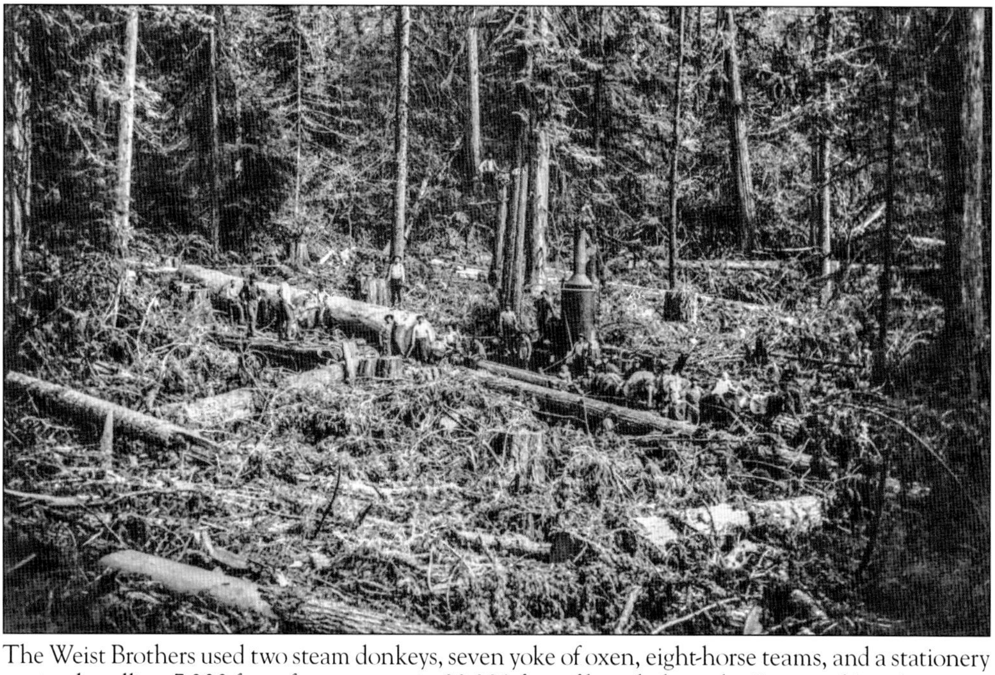

The Weist Brothers used two steam donkeys, seven yoke of oxen, eight-horse teams, and a stationery engine handling 7,000 feet of rope to put in 80,000 feet of logs daily in the Promised Land country at the end of the 20th century. Horse loggers in East County included J.B. Haynes on the Mox Chuck Slough. W.D. Mack was in the same area in 1891, and Polson in 1895. (PM.)

This worker is bringing in a load of cants via a skid road. They will be sold as railroad ties and dock cribbing. Skid roads were built with small logs, six-by-one-foot in length, placed eight feet apart. When a camp was logged out, skid roads were left to the elements. (AMH.)

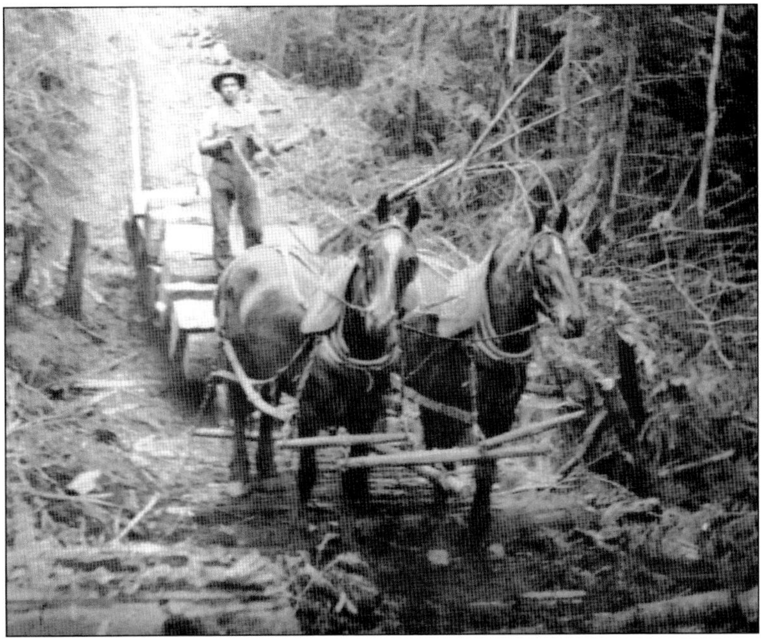

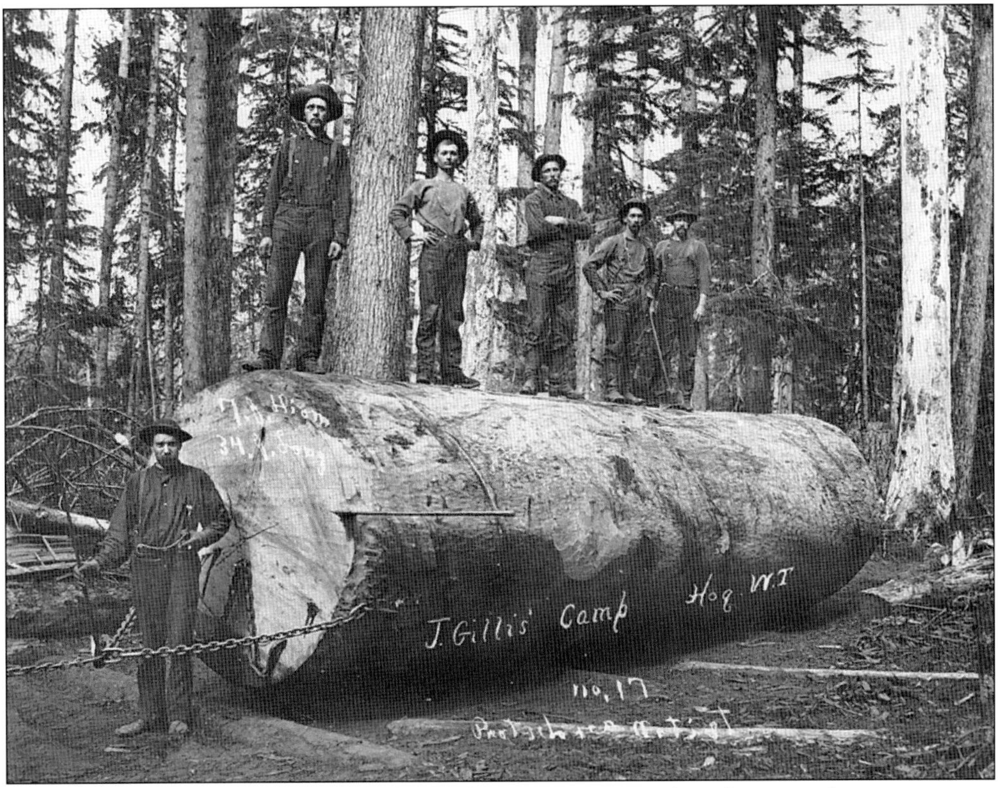

Here, the J. Gillis ox teams are shown bringing in nine-foot-diameter logs measuring 286 feet in length. According to Alex Polson, it took a bellowed "h'ist" to get the team moving, along with a good smack from a "go devil" (goad) and some dark blue language. (Photograph by Tollman & Pratch; AMH.)

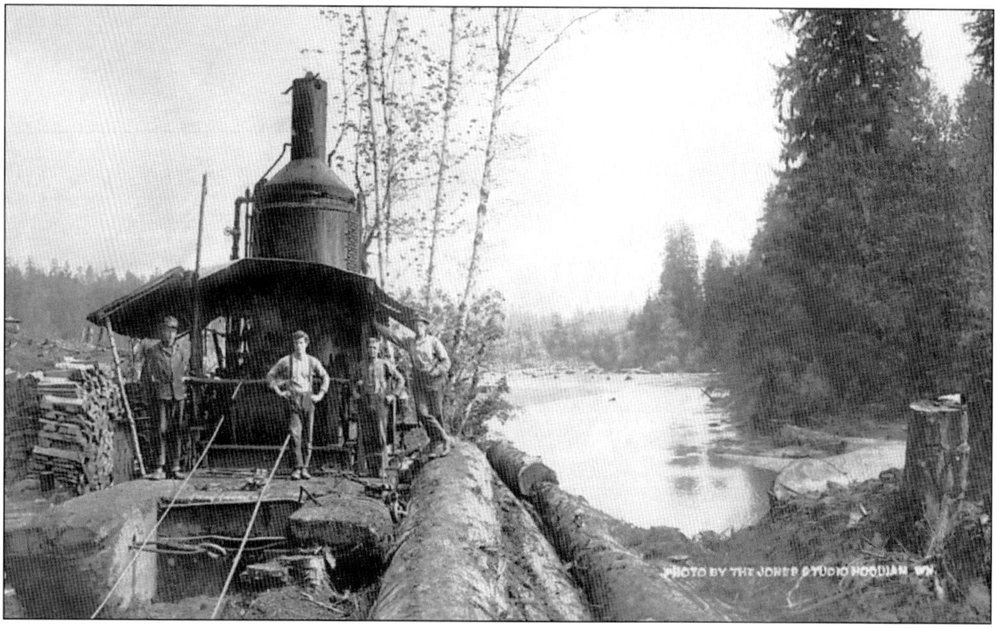

The Walker brothers, E.R., John, Jeremiah, and Herman, formed the Walker Timber Company. They are shown here working with their donkey on the Humptulips River slough, close to the same time as the Soljeth & Moe outfit operated. Their close neighbors, Ellison & Egge Logging Company, worked out of the Promised Land in 1883. (AMH.)

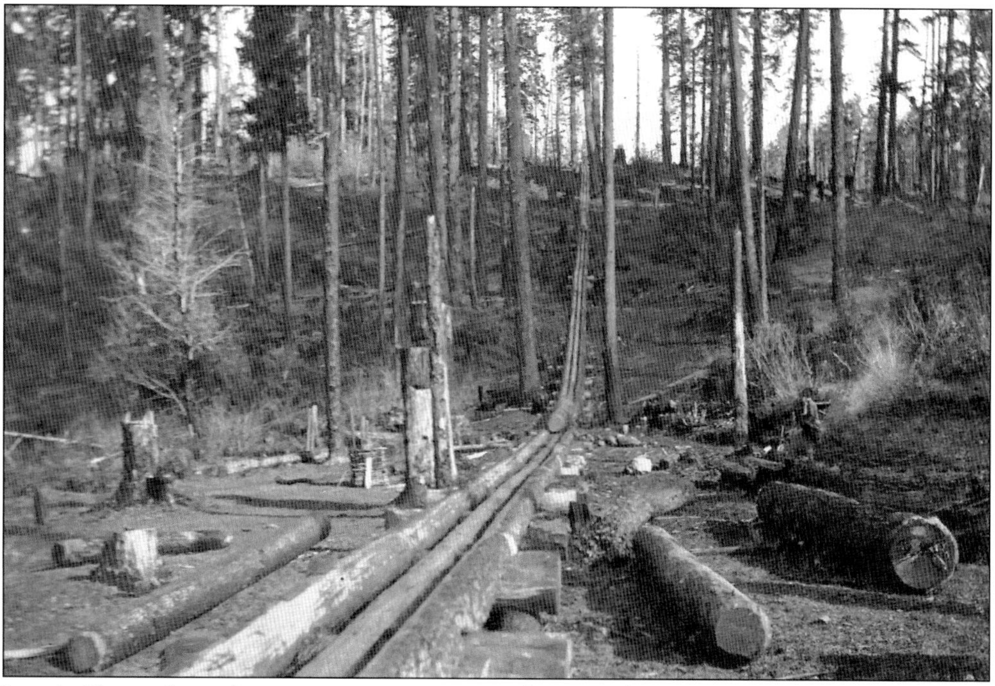

Log skidding pole roads were an important part of oxen and early steam donkey logging. Note how parallel side logs formed a shoot for skidding log. The name Skid Road (not Skid Row) was later applied to the areas of logging towns where streets were lined with saloons, houses of ill repute, and logger hotels. (AMH.)

Alex Polson was proud of this photograph displaying the size of logs taken out by the beginning Polson Logging Company crew. It is almost beyond imagining a log so large that an upright horse and rider could stand in the butt-rot hole of the fallen tree. (PM.)

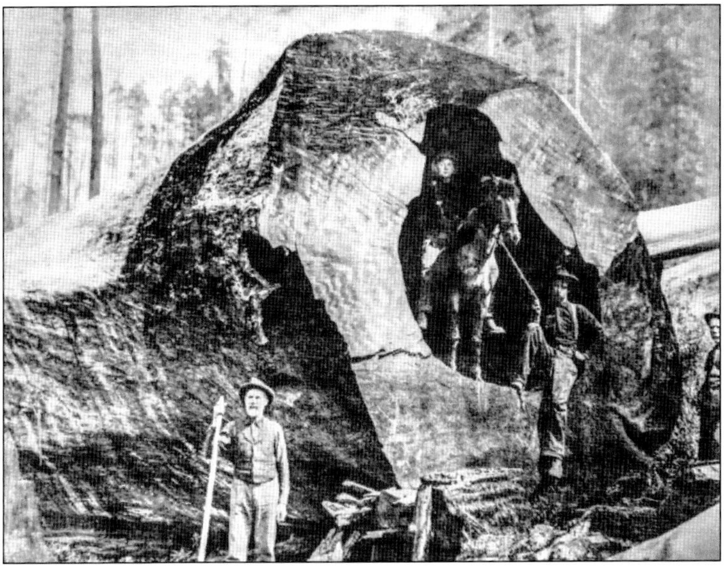

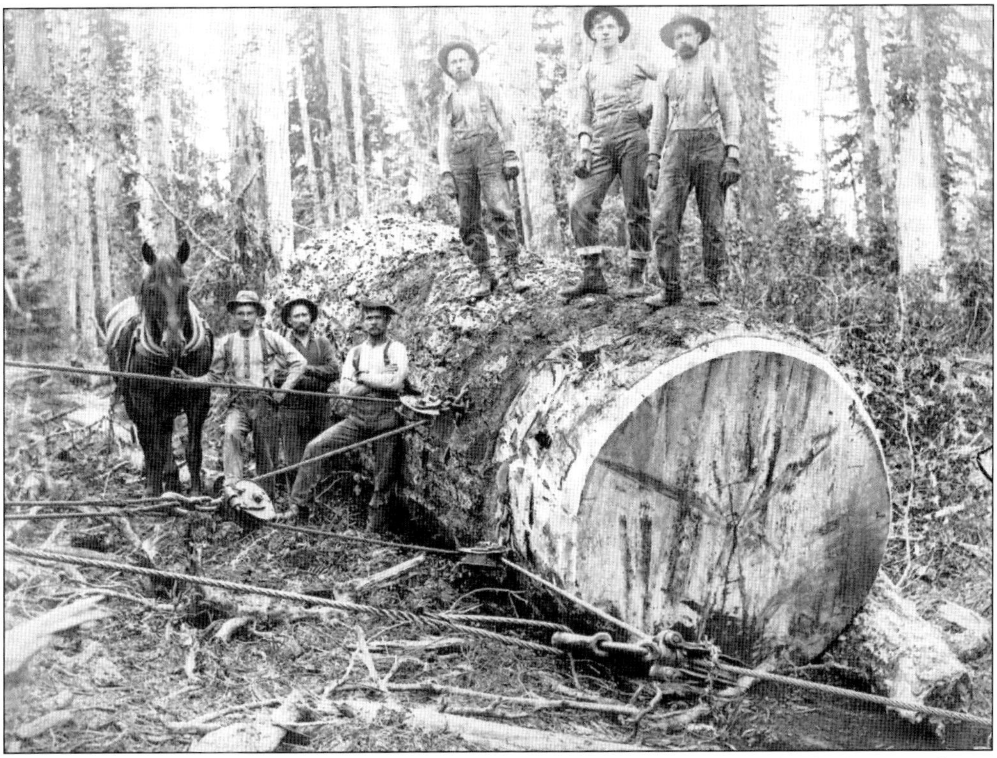

When the Dolbeer donkey spool engine first appeared in early 1887 on the harbor, one horse pulled logs to the roll dam or splash dam landings. The rigging in this photograph shows that the early loggers understood physics and engineering. By the late 1890s, the Dolbeer was gone from the woods. (Photograph by H.G. Nelson; PM.)

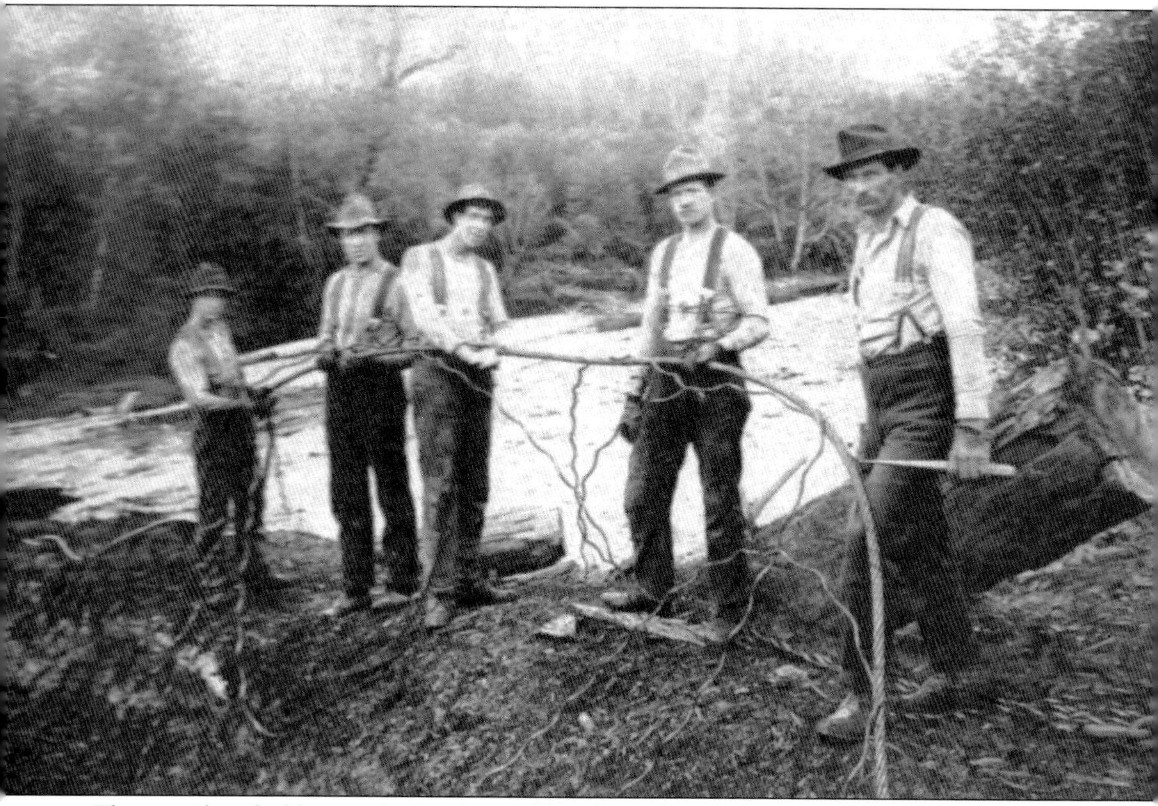

These unidentified loggers display their cable-splicing knowledge, a necessary skill in the field. Metal splicing spikes were fashioned in the same manner as the old mariner's marlin spike (fids) used to splice rope. Later, Jim Mason was known as a virtuoso who could make the wire sing. He had taken splicing from a skill to an art form. (CHSGWC.)

Three

Steel Comes to the Woods

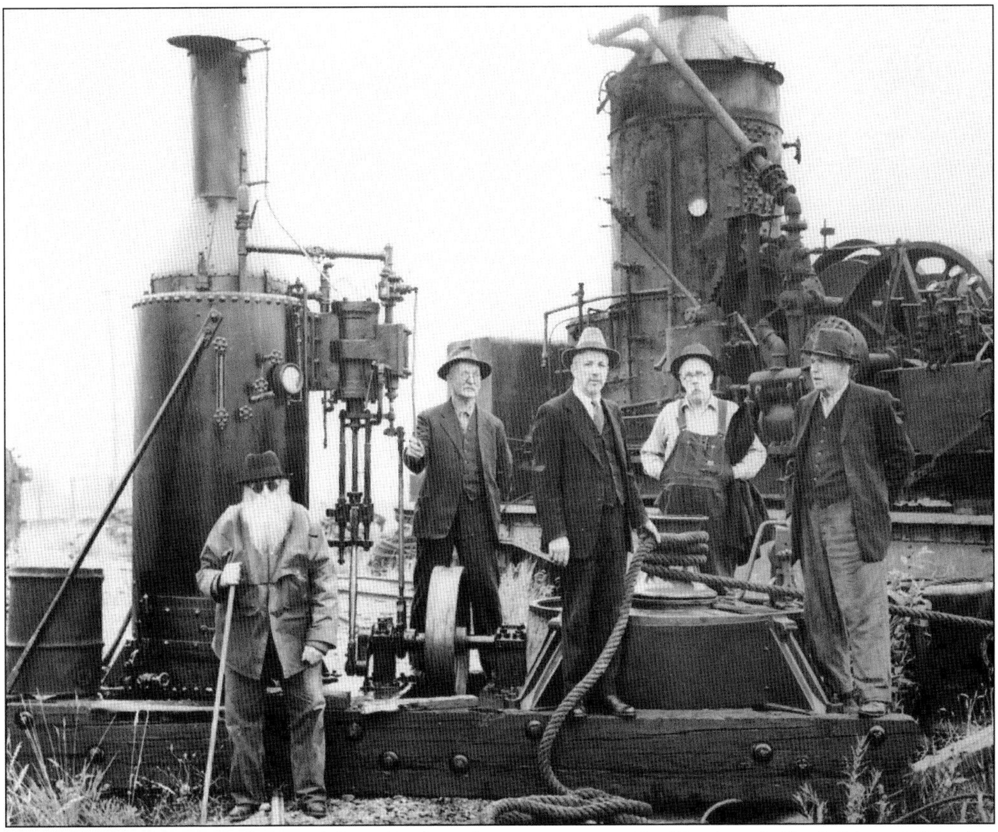

The famous oxen puncher and first donkey drum engineer, Cy Blackwell (far left) and Charles Neeson (second from right), pose with three unidentified men in front of the first two steam donkeys of their kind on the harbor. The old Dolbeer with its ship's capstan (left) is quite a contrast to the new spool drum steam donkey (right) brought in to log near the Humptulips River. (AMH.)

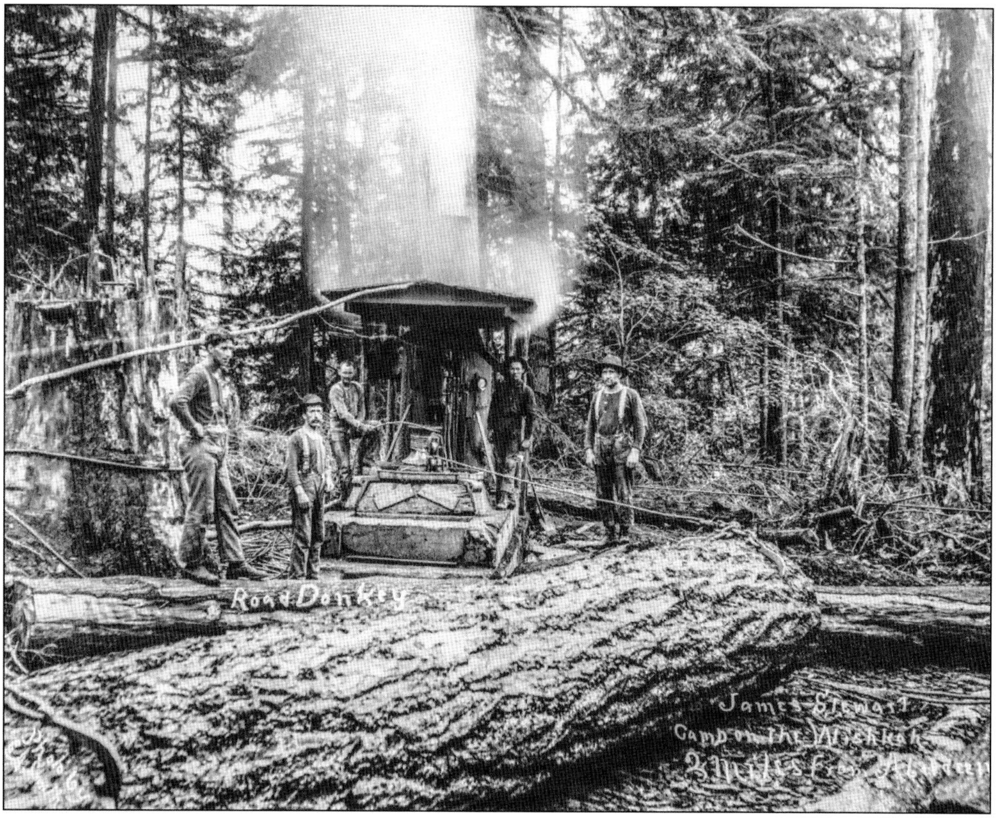

James Stewart's road donkey was an exciting addition to his first camp on the Wishkah River, located two miles east of Montesano in the 1880s. He called a photographer to record his "Roader" wide-faced donkeys with two drums that could be moved to and from logging sites. (PM.)

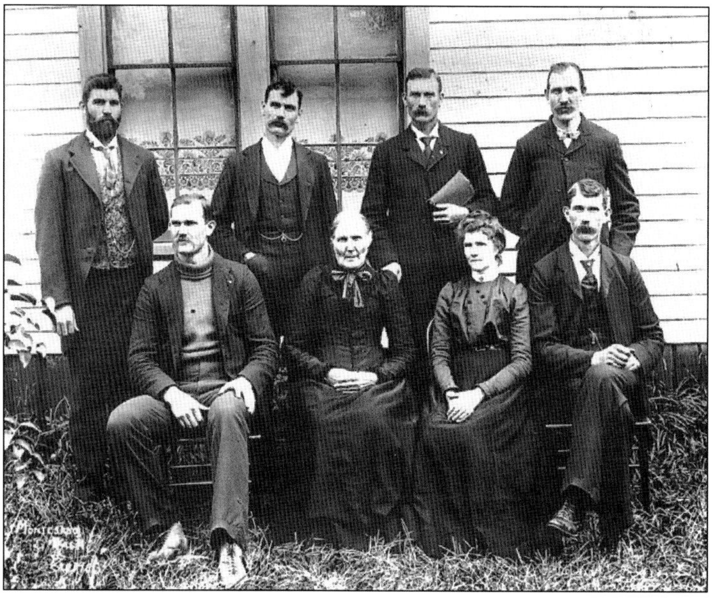

The Caldwell boys, who formed one of the biggest logging outfits in the county, were known for their size and their logging prowess. It was said that when they walked to church with their mother, Frances, in Montesano, the boardwalk shook from end to end. Shown here are, from left to right, (first row) George, Frances, Lucinda "Lucy" Jane, and Dick; (second row) Byron, Oliver, William "Preacher Bill," and Henry. (AMH.)

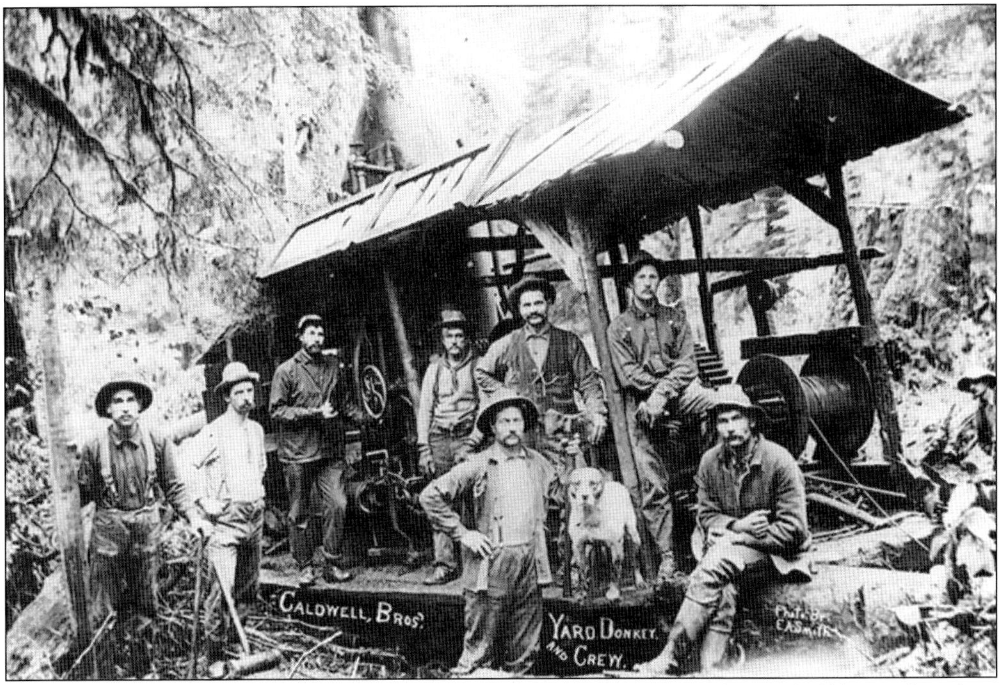

The Caldwell Brothers crew poses for this E.A. Smith photograph with their bulldog and yarding donkey in the John's River area at the turn of the 20th century. The South Beach area of the harbor was tricky to log, since so much of the lowlands were swamp areas. (AMH.)

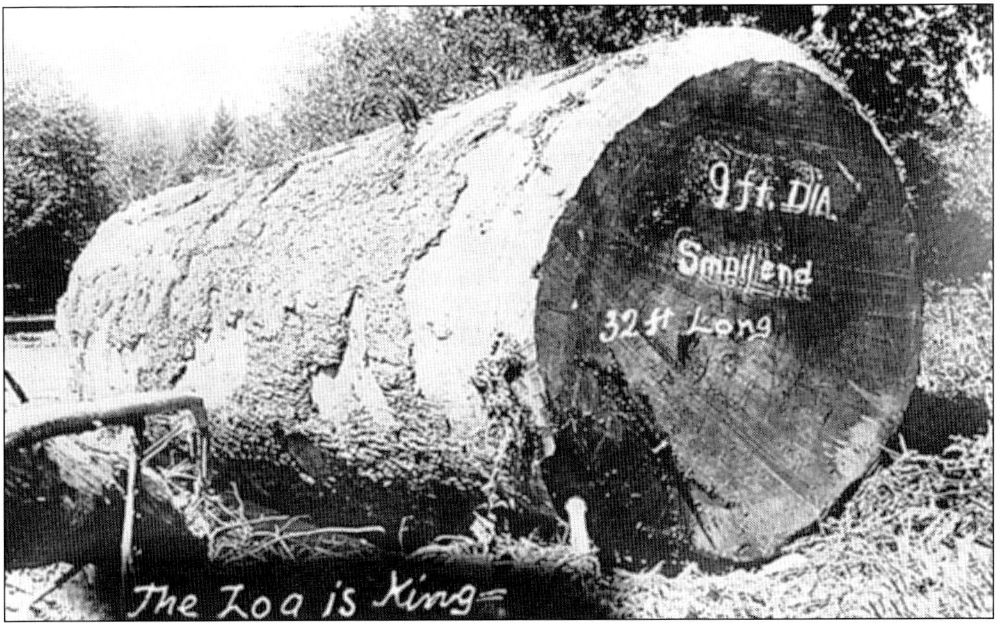

The log is king. This 32-foot-long spruce log had a nine-foot butt. On April 15, 1926, Suitico (Sunset Lumber Co.) Camp 9 harvested one of the largest logs ever taken out of North River's Elk Creek area, measuring 100 board feet on the top end and 122 board feet on the butt. The camp was bringing in "green gold," with two side camps shipping out, from stump to cars, 957 rail loads in a sort time span. (AMH.)

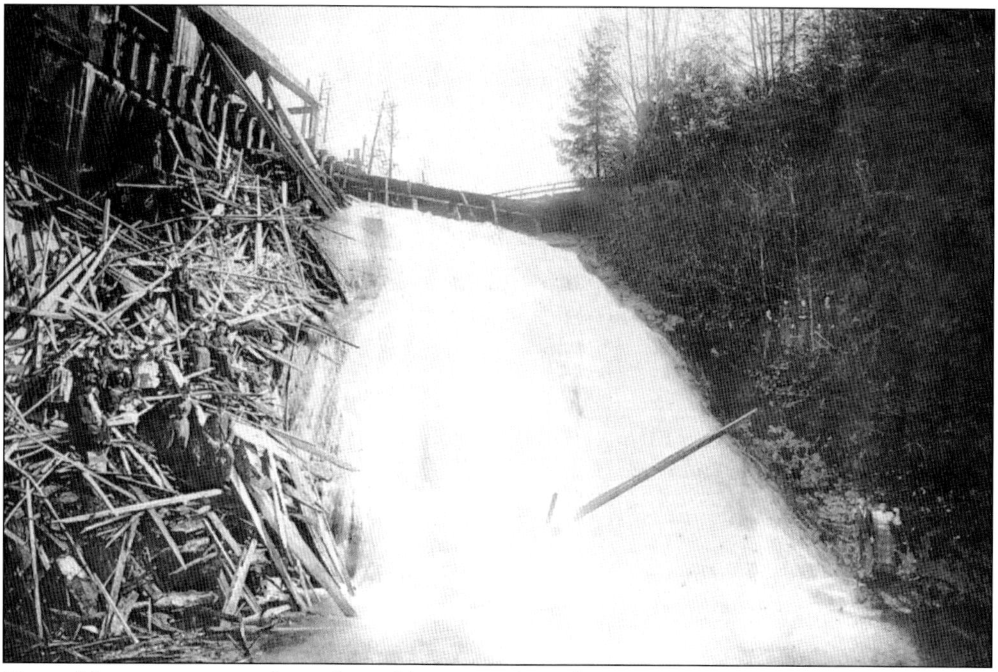

Ground logging was not the only problem in early endeavors; once cut into lumber, anything could happen before the product made it to market. When the flume at the Lake Sylvia Mill at Montesano blew out, sending lumber downstream, months of production was lost. The mill began production in 1870 with an up-and-down, single-blade, vertical slash saw. (AMH.)

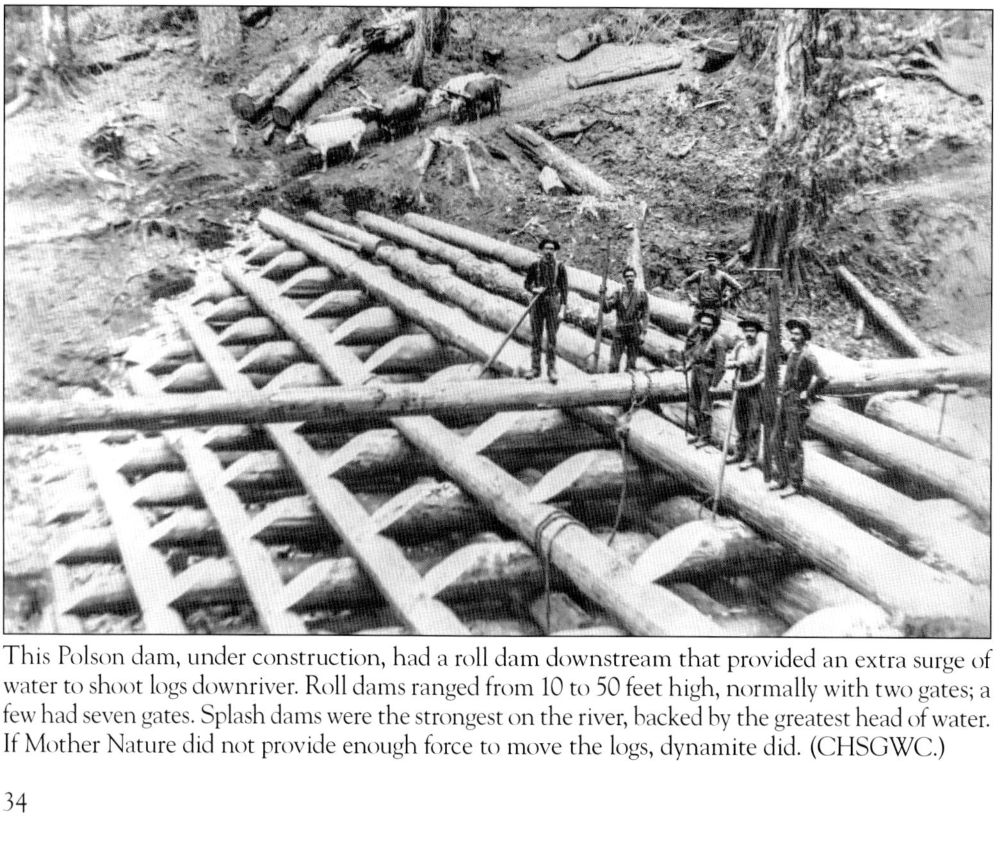

This Polson dam, under construction, had a roll dam downstream that provided an extra surge of water to shoot logs downriver. Roll dams ranged from 10 to 50 feet high, normally with two gates; a few had seven gates. Splash dams were the strongest on the river, backed by the greatest head of water. If Mother Nature did not provide enough force to move the logs, dynamite did. (CHSGWC.)

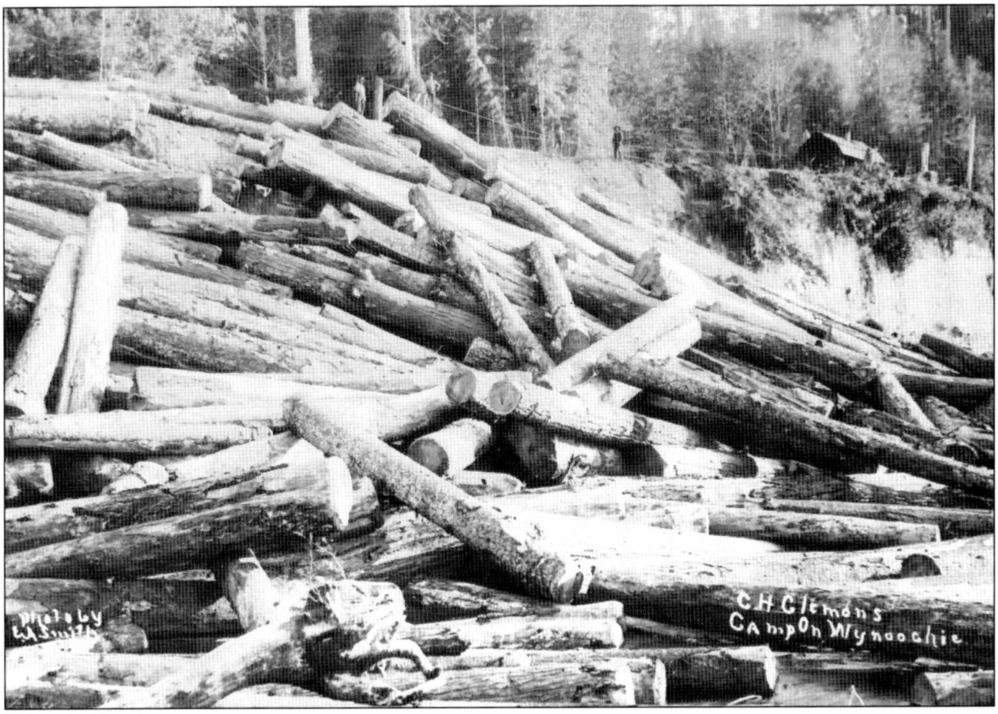

This A.C. Clemons jam is merely logs against the bank, waiting to be flushed downriver by a Winkleman splash dam, whose designer was considered one of the finest builders. Ben Turner, another quality builder, supplied dams to the Charles Peterson Company on Charlie Creek and Clyde Brothers on Big Creek. Nels Esses concentrated on the Promised Land area. (CHSGWC.)

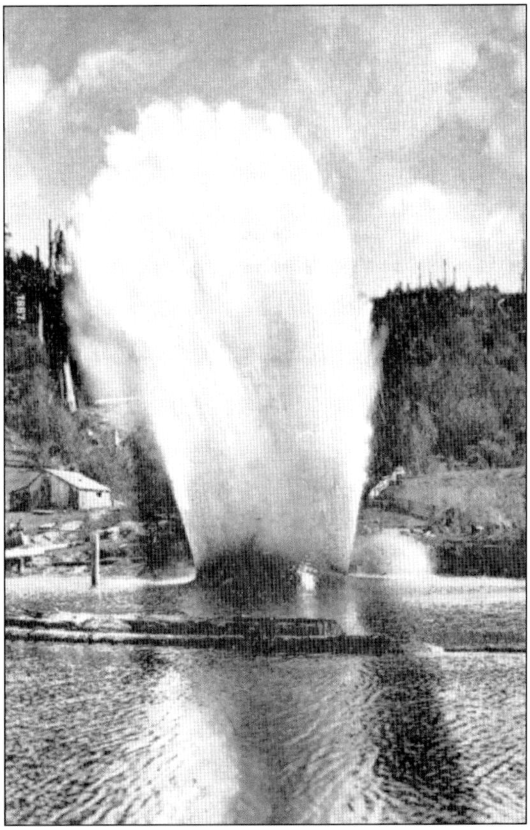

This "going to splash" event occurred on an early Polson dam, located on present-day Grand Street in Hoquiam. This dam was basically a roll dam that flushed logs into the Hoquiam River for a short trip down to the mills on the bay. The Frank Stenzel Logging Company put 12,000 feet of logs into the Humptulips River in 1899 near Deep Creek—all through hand logging. His greatest and most ingeniously installed splash dam covered 200 acres of logging lands. Around the same time, splash dams were installed on Chenois Creek to open log storage on Jessie Slough. (CHSGWC.)

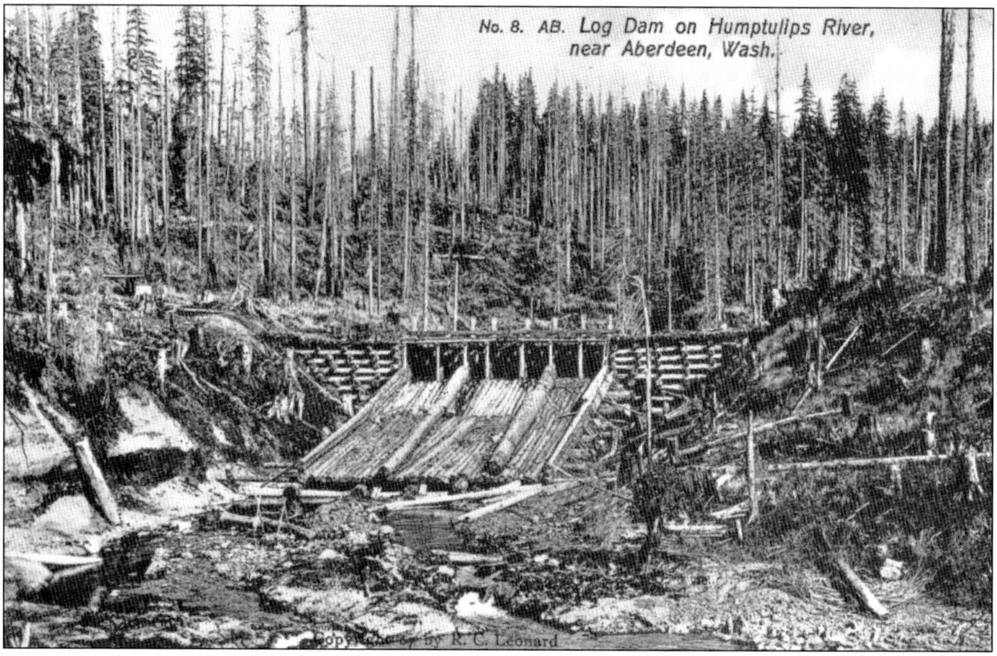

This dam, one of 27 splash dams on the Humptulips River, sent logs down a 34-mile stretch to the slough of the North Bay. The widespread use of storage and rafting areas for logs created an entire new industry of logging companies. A Humptulips River logging outfit, Cameron-Hoover and Corkery, invested in a number of dams that cost $25,000 to $30,000 to construct. (CHSGWC.)

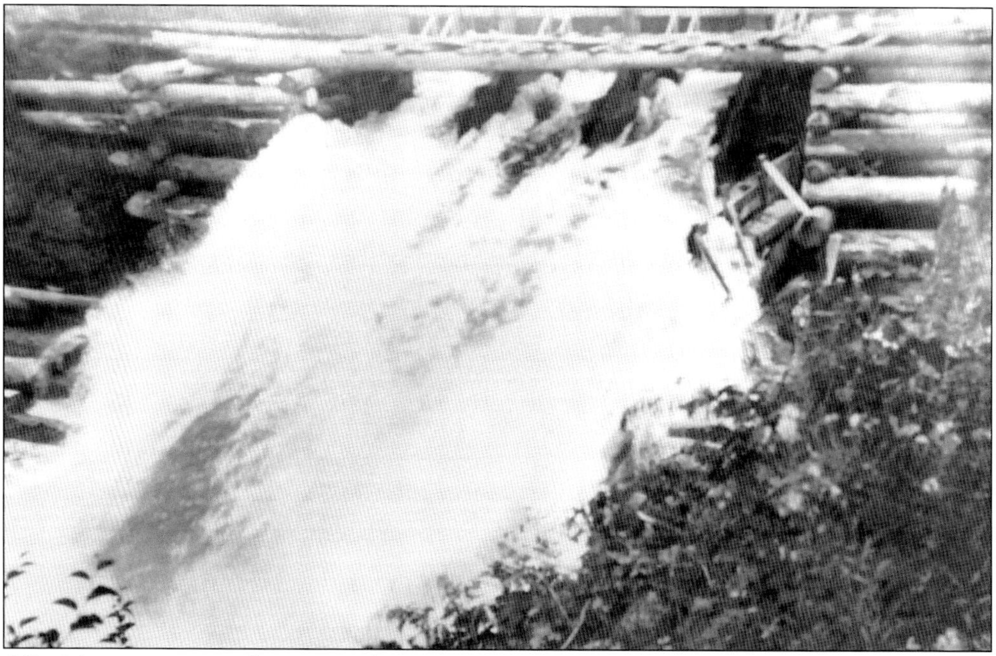

All of the gates were wide open on this splash dam. Freshlet water cascaded downriver, carrying thousands of board feet of timber to the booming grounds. Each spring, dams were used to move millions of feet of winter logging efforts. Splash dams located on streams with sufficient flows could also be used to store logs, to be sent downstream at more convenient times. (CHSGWC.)

Four

Whistle Punks and Steam Donkeys

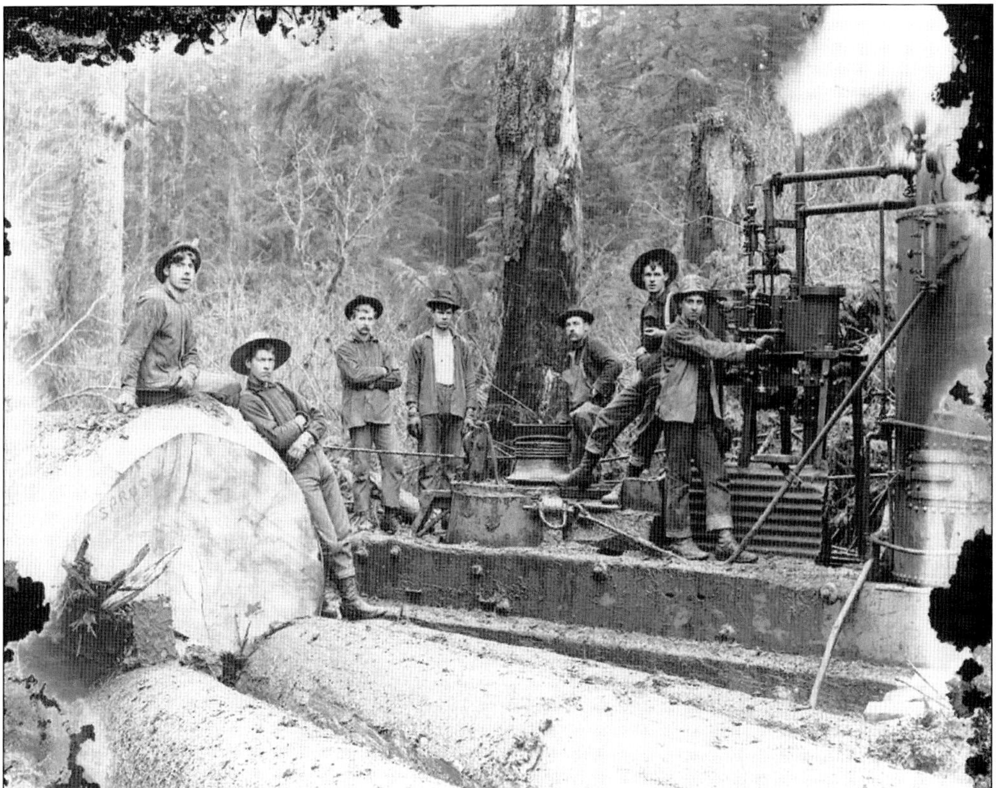

The coming of steam began the colorful era of logging. It was now a culture, not just an extra job position for homesteaders. Steam-era logging took brains, brawn, guts, and a whole new vocabulary of cusswords. This first Dolbeer on the harbor, brought in by Cy Blackwell, hauls in a sniped log. (AMH.)

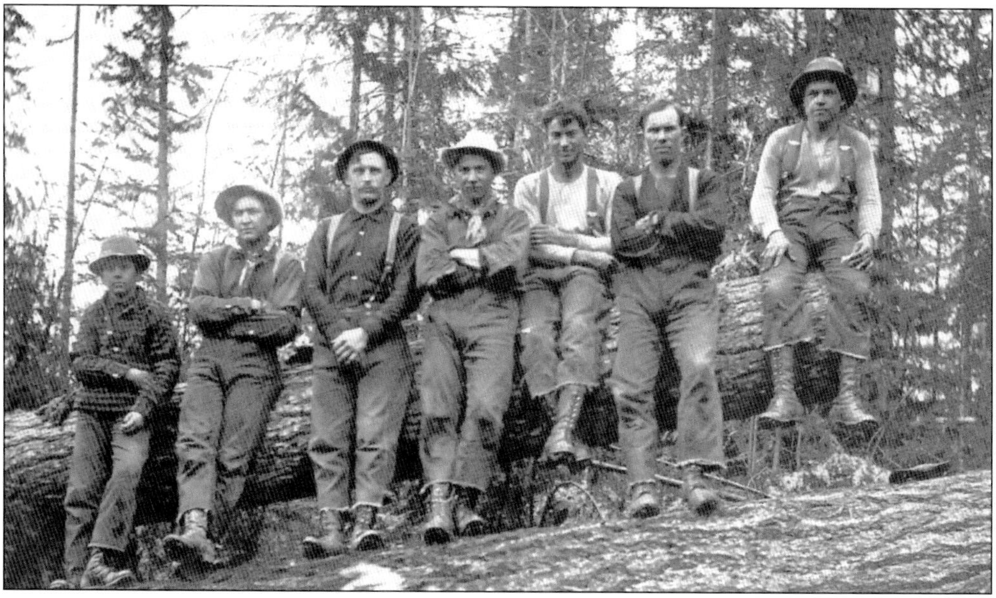

Steam allowed 15-year-olds, like the young whistle punk at far left, to proudly pose with "real loggers." They are wearing "tin" (parafined or linseed oiled rainproofed), staged off pants with legs cutoff at the boot tops to prevent snagging on brush and limbs, ubiquitous gray underwear, suspenders, corks (caulked boots), and felt hats. A whistle punk pulled the wire attached to the donkey engine whistle to signal the engineer that a log was ready for landing. (CHSGWC.)

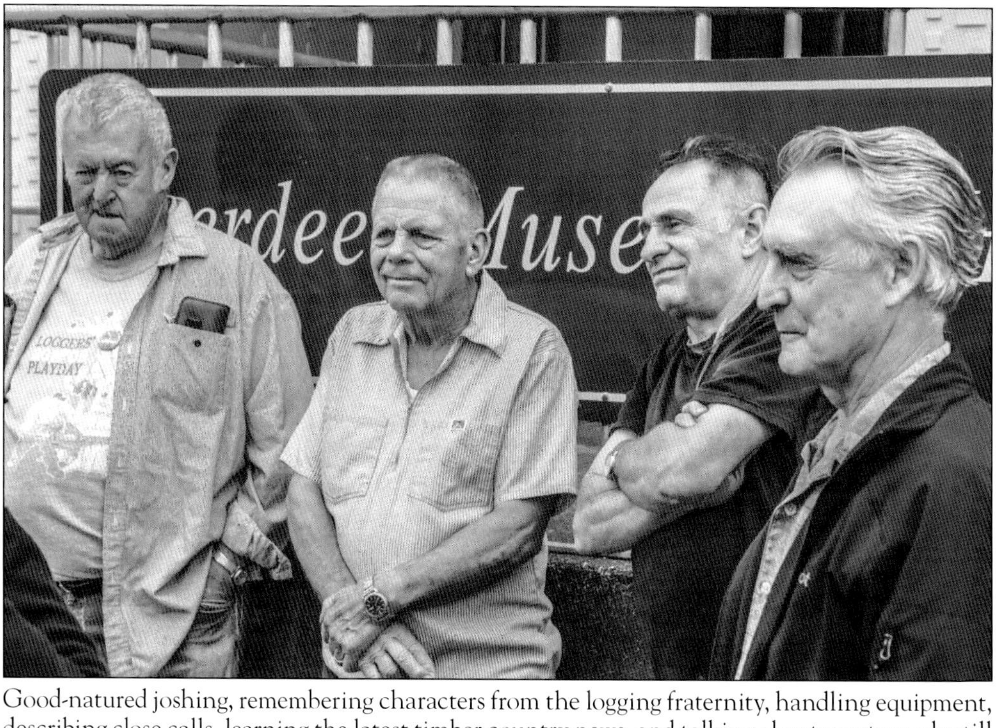

Good-natured joshing, remembering characters from the logging fraternity, handling equipment, describing close calls, learning the latest timber country news, and talking about great people still provide a good time for retired loggers. Shown here are, from left to right, Guy DuGay, Howard Brunstad, Robert Sholes, and Dan Schessler. (Brian Woodwick.)

This image shows the guys waiting for supper at Rayonier Camp 6 near Humptulips in the early 1960s. It was a good time to sit in the late afternoon sun, yaking away to catch up on logs, loads, new equipment, and what some jackass did during the day. (CHSGWC.)

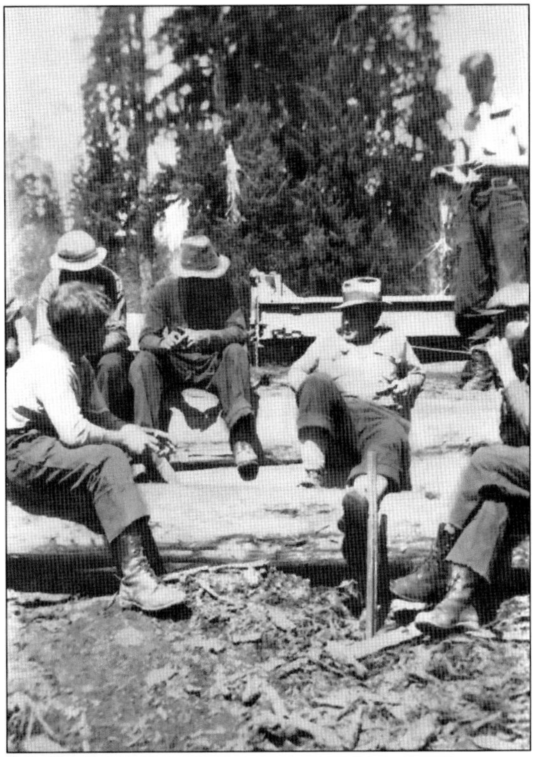

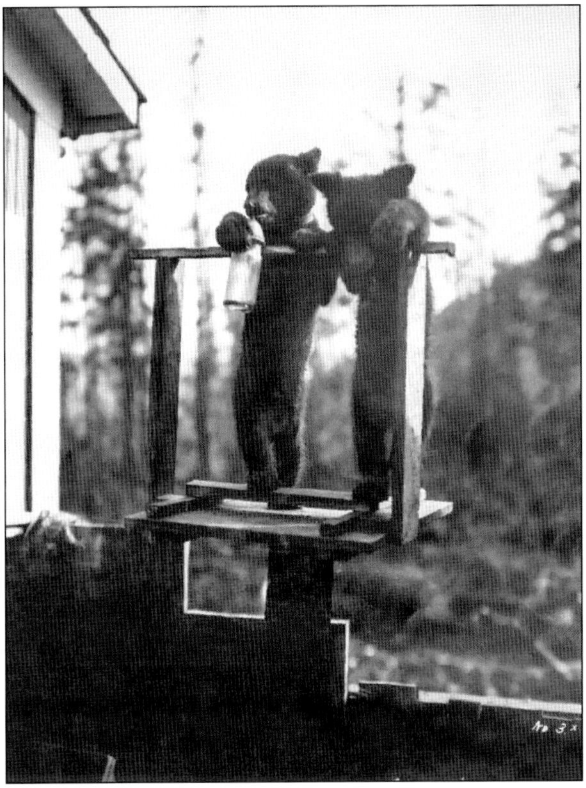

Loggers worked hard and played hard. It was not just hunting and fishing on days off. Sometimes, going into town was not worth the effort. Playing with animals was a favored pastime, and many camps kept bear cubs as pets. In some cases, the mother bear had been killed by a falling tree. (AMH.)

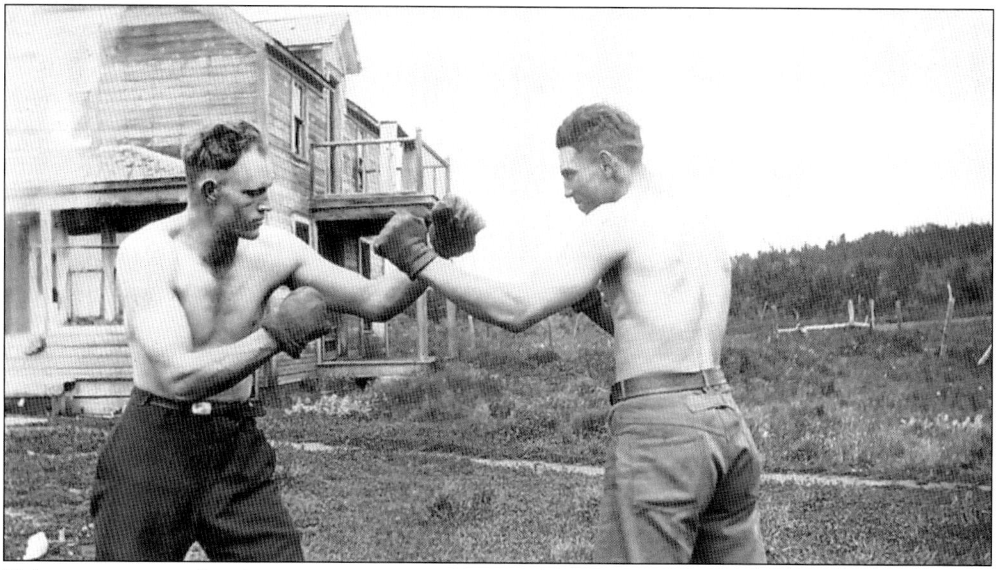

Fighting was considered an art, not an expression of anger, in camps or in town. Loggers might beat up on each other before going off arm-in-arm, laughing and still being good friends. Just as cowboys had gunslingers to challenge, loggers had the muscular tree climbers to challenge on a regular basis. (CHSGWC.)

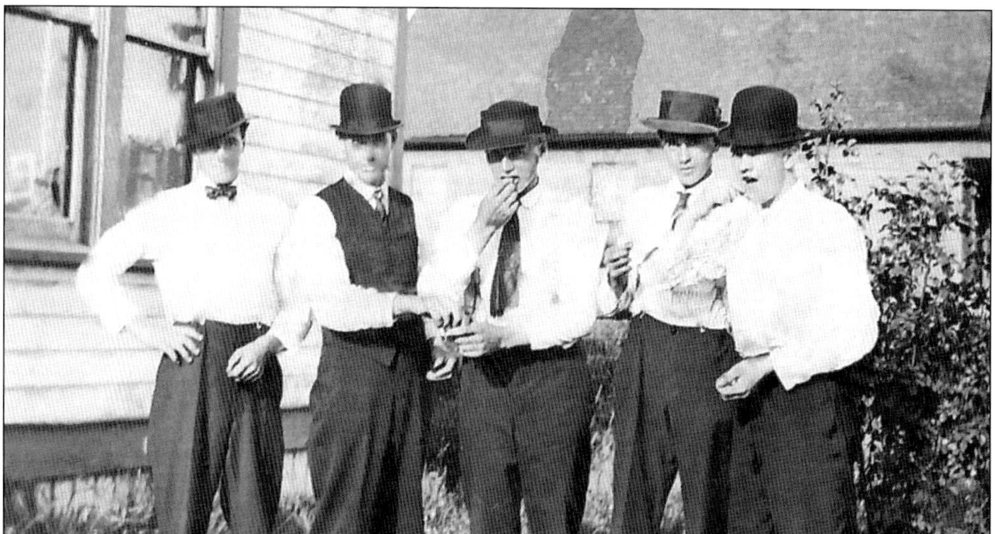

A snooze-chewing contest between these dressed-up dudes would have warmed the hearts of taxmen. In one year, the government collected taxes on 40,655,395 pounds of snuff—mostly Copenhagen, although there were 20 kinds available. Today, the faded snooze-can ring on back pockets is rarely seen. (CHSGWC.)

Boozing in this unnamed Grays Harbor saloon provided a fine picture for the postcard photographer, who roamed the town at all hours. A night on the town was not complete without at least one trip to the famed Humbolt Saloon in Aberdeen. The area contained more saloons than any place in the state. (CHSGWC.)

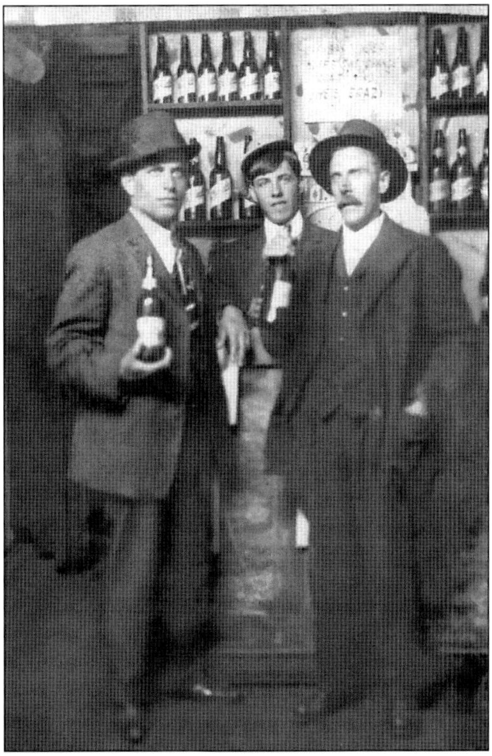

Loggers who made it back to camp on a "Blue Monday" following a "blow out" in town often returned with a postcard to send home. In 1914, Lester Lewis posed by a bright, shiny moon to send this card to his mom. Real-photo postcards were also the preferred way to correspond with the girlfriend of the month. (CHSGWC.)

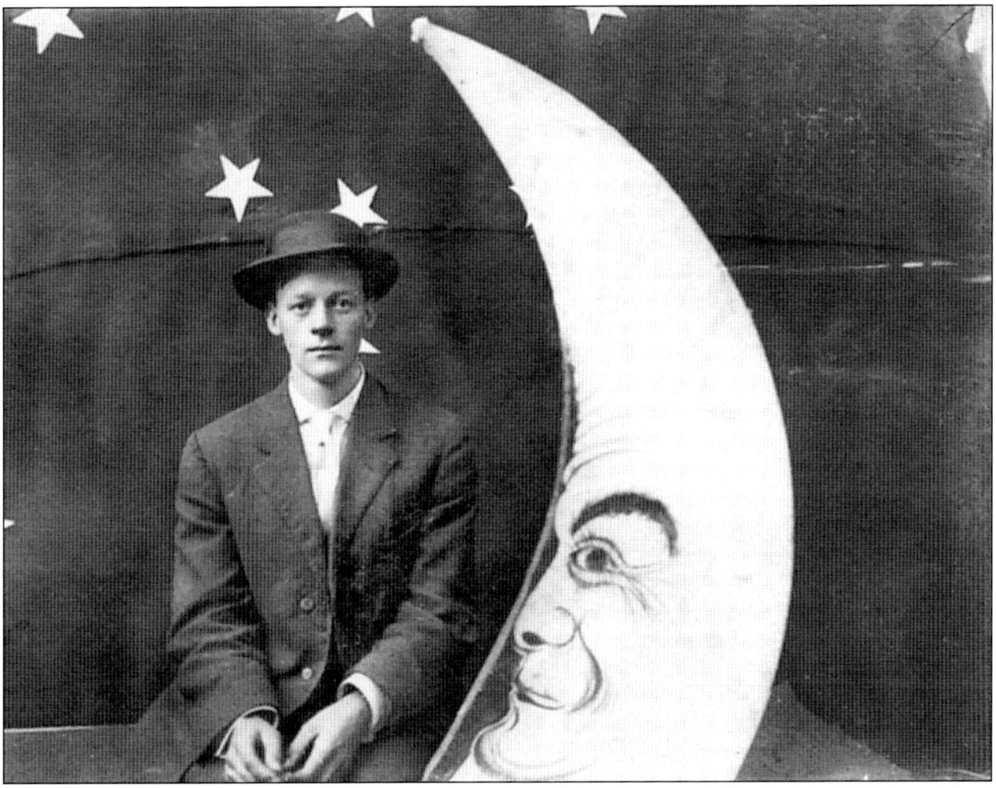

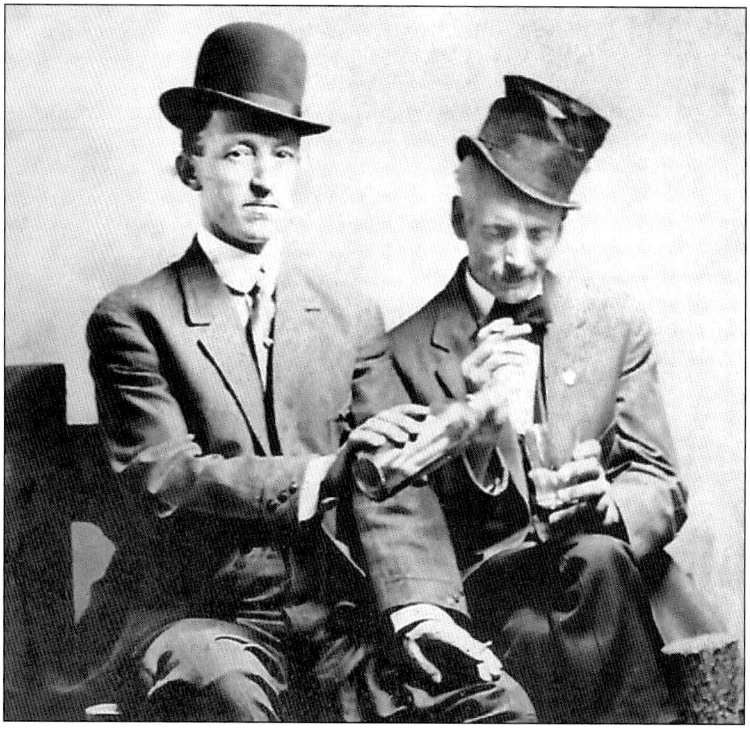

Posing drunk for a real-photo postcard was considered high humor to mail to a friend. Logger humor abounded in many forms, and drinking was considered fun as well as a reprieve from the unremitting hard work in the camps. For the most part, drinking occurred in town, as it was too dangerous to mix booze with falling trees. (CHSGWC.)

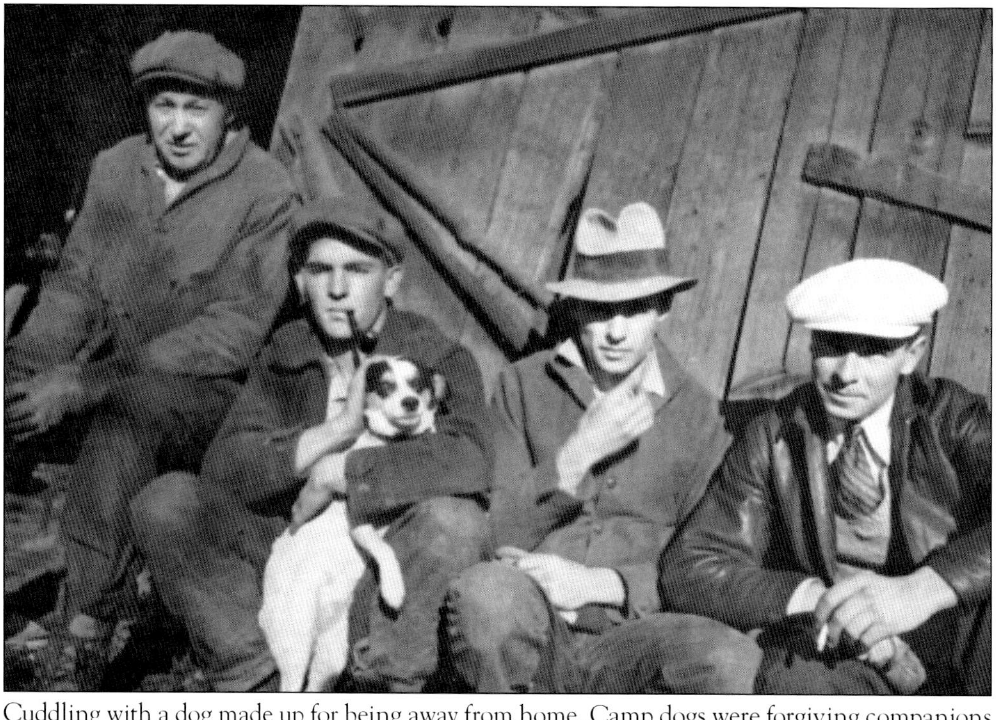

Cuddling with a dog made up for being away from home. Camp dogs were forgiving companions to the younger loggers, bear chasers at the camp cooks' garbage dumps, and listening ears for the bed-makers during daylight hours. They provided a topic for conversation and an opportunity to josh each other over ownership. (CHSGWC.)

While it was not discussed in polite company, houses of ill repute were of economic benefit to the towns surrounded by camps. This log cabin crib was located in the Humptulips area in the late 1890s. The proliferation of such businesses in the Aberdeen and Hoquiam areas during World War II caused the military to list the towns as off-limits to servicemen. (CHSGWC.)

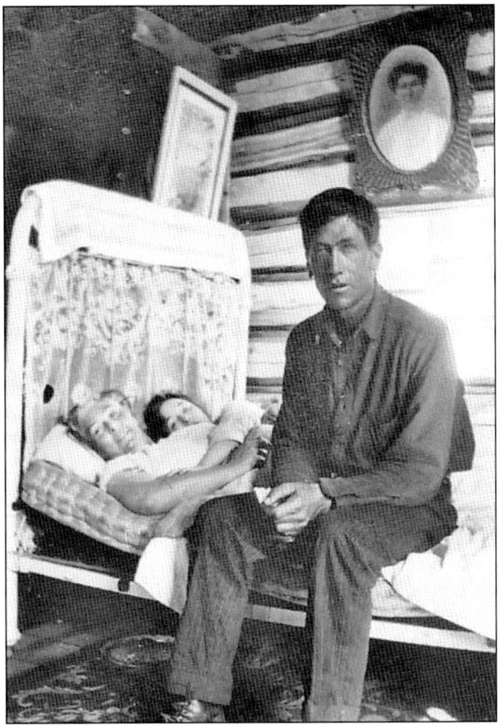

Dale G. Hill went to work in 1963 at Rayonier's Camp 14 as a bed-maker and cook flunkey. He was a highly creative individual. He died in 1974, leaving behind a legacy of rare camp life photography and taped conversations of logging history for his family. (DHC.)

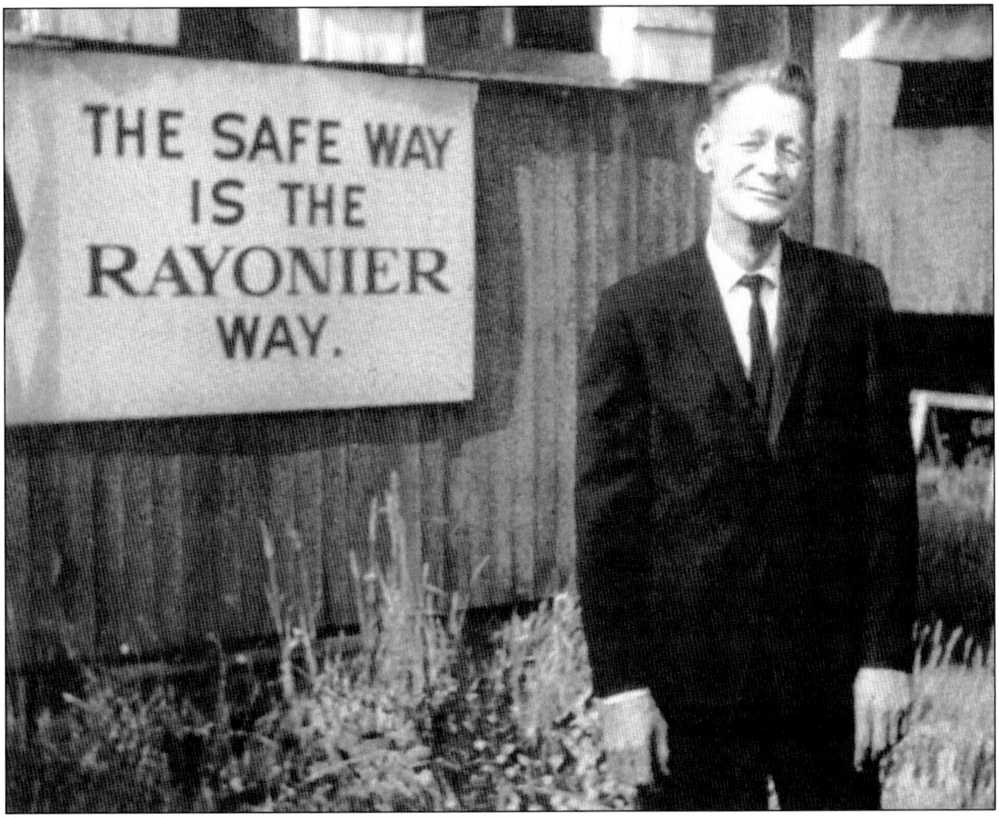

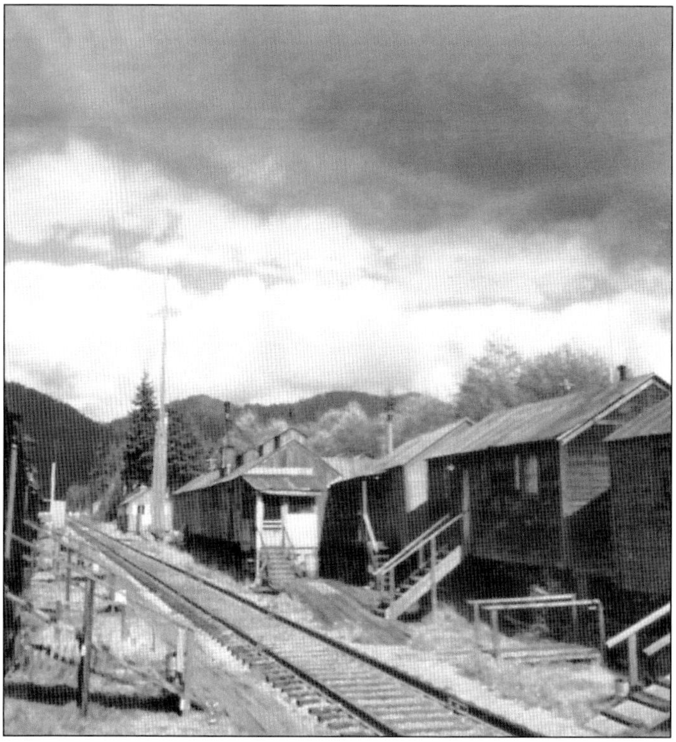

This photograph is looking down the main line of Camp 14 in October 1964. When moved by Polson Company to its location in 1936, the camp was already older than the second-growth timber surrounding it. Home to 250 men, in 1965, it was the last remaining camp on wheels in the nation. (DHC.)

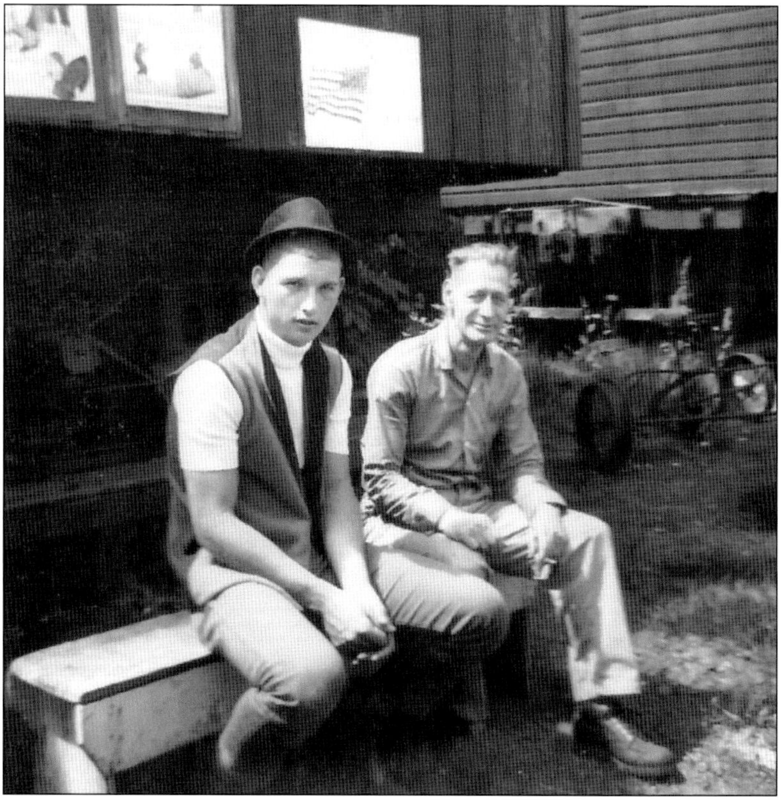

Ira H. Thomas (left) and Dale Hill shared a great friendship of jawboning, music, and target practice with a pistol or rifle. Thomas was killed in a yarding accident in November 1971 while working as a rigging slinger at the Rayonier Crane Creek Reload Camp, located 45 miles north of the New London log dump. (DHC.)

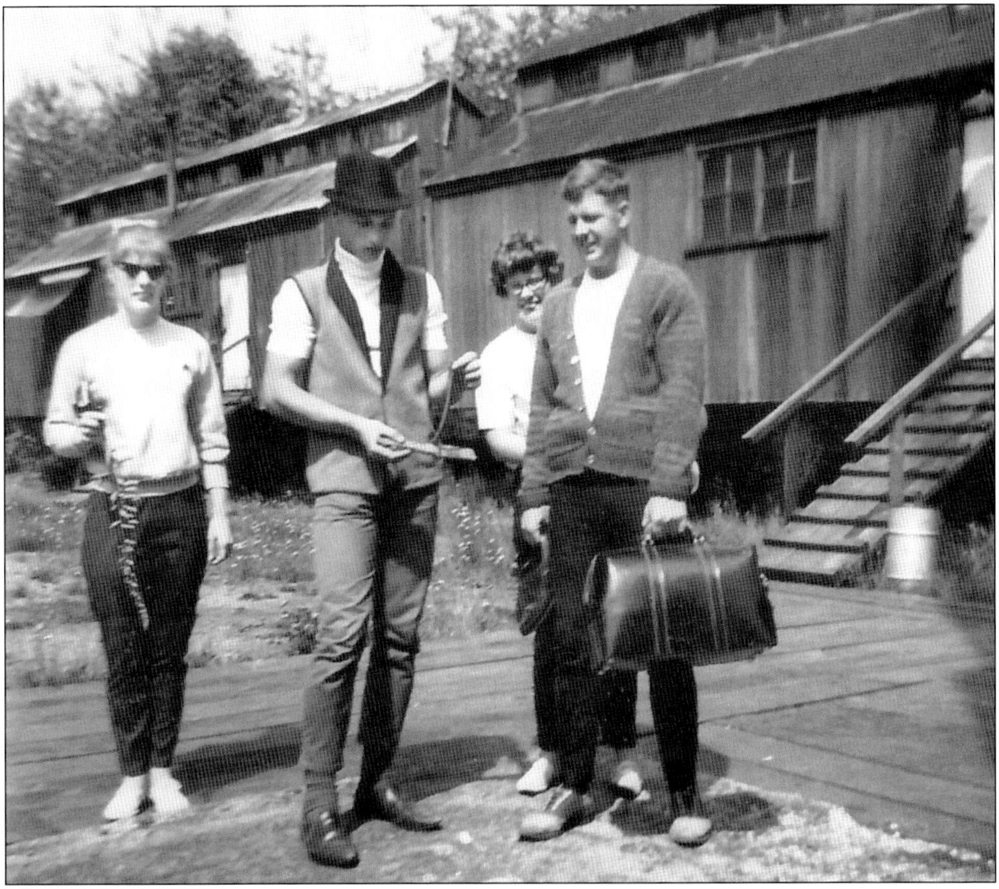

Shown here are, from left to right, Esther Hill (Kinney), Ira Thomas, Shirley Hill, and Frank Hill. They are preparing to leave camp after a visit with Esther, Shirley, and Frank's dad, Dale Hill. Dale loved the visits more than he liked going to Gambar's logging café in Hoquiam. Gambars served as a meeting place for loggers until the early 2000s. (DHC.)

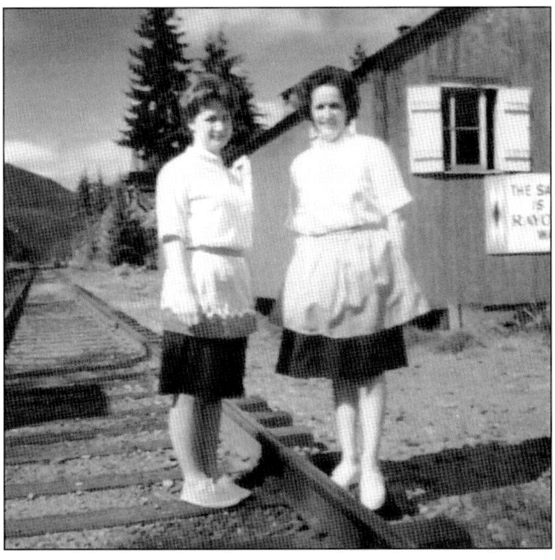

Ray Anne Taylor (left); her dad, "Hap" (not shown); and Helen Nation were cooks at Camp 14. Women sure beat the old-time male cooks in the looks department! A cook's day began at 3:00 a.m., with firing up the gas ranges. The day involved fixing breakfast; packing lunches; making bread, cakes, pies, and cookies; and getting dinner. The day ended at 10:00 p.m., with the dishes all done. (DHC.)

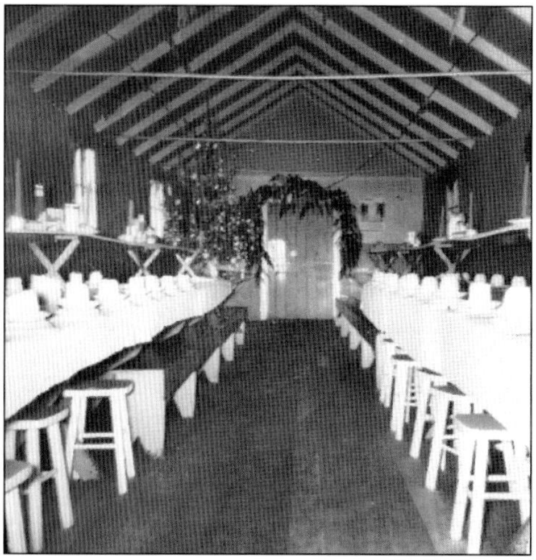

The cookhouse/recreation hall was truly home to the guys. It was the place to fill up on mounds of home-cooked food, kibitz, read mail or hunting magazines, and enjoy a holiday meal with their own decorated tree straight from the woods. (DHC.)

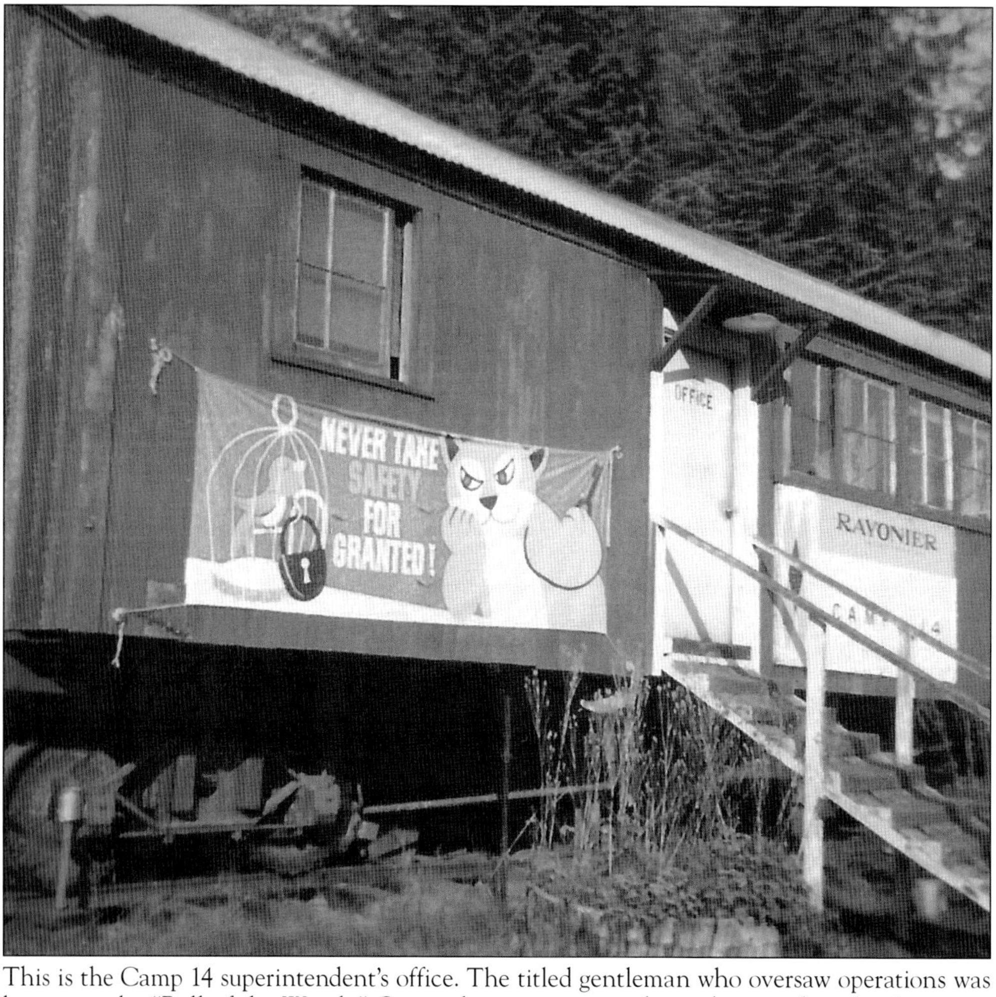

This is the Camp 14 superintendent's office. The titled gentleman who oversaw operations was known as the "Bull of the Woods." On payday, it was a popular gathering place for the crews, slicked up and waiting to head into town with a buddy or on the bus. (DHC.)

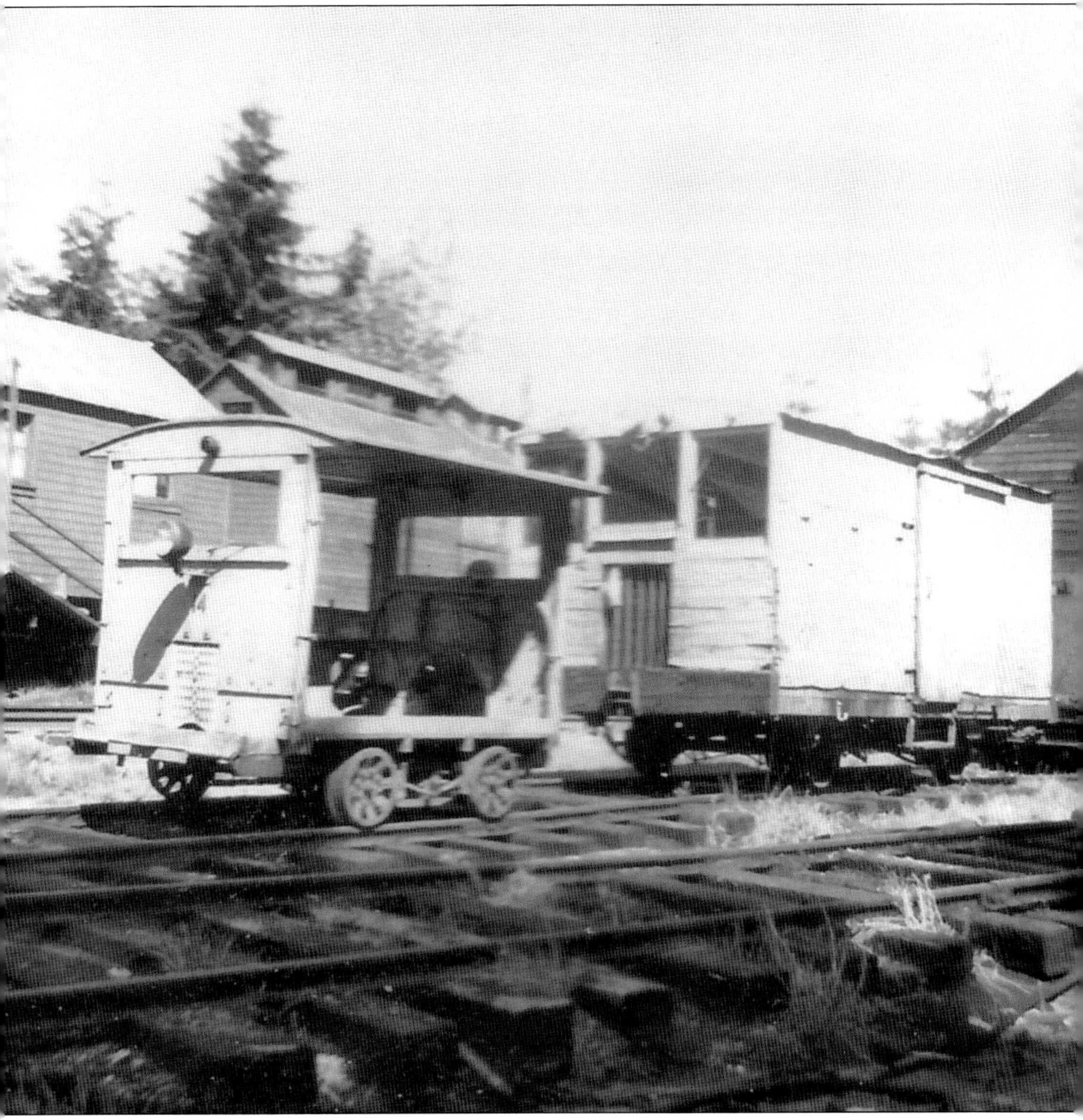

This Rayonier railroad speeder (left) and the rail "crummy" (right) carried men and supplies to keep the line and various show sites in repair. The remodeled, smaller crummy was referred to as the "outhouse on wheels." Speeders and any rig used to perform odd jobs was called a jillaghike, a gillagahike, or a yellagahike, according to how many Scandinavians were in camp. (DHC.)

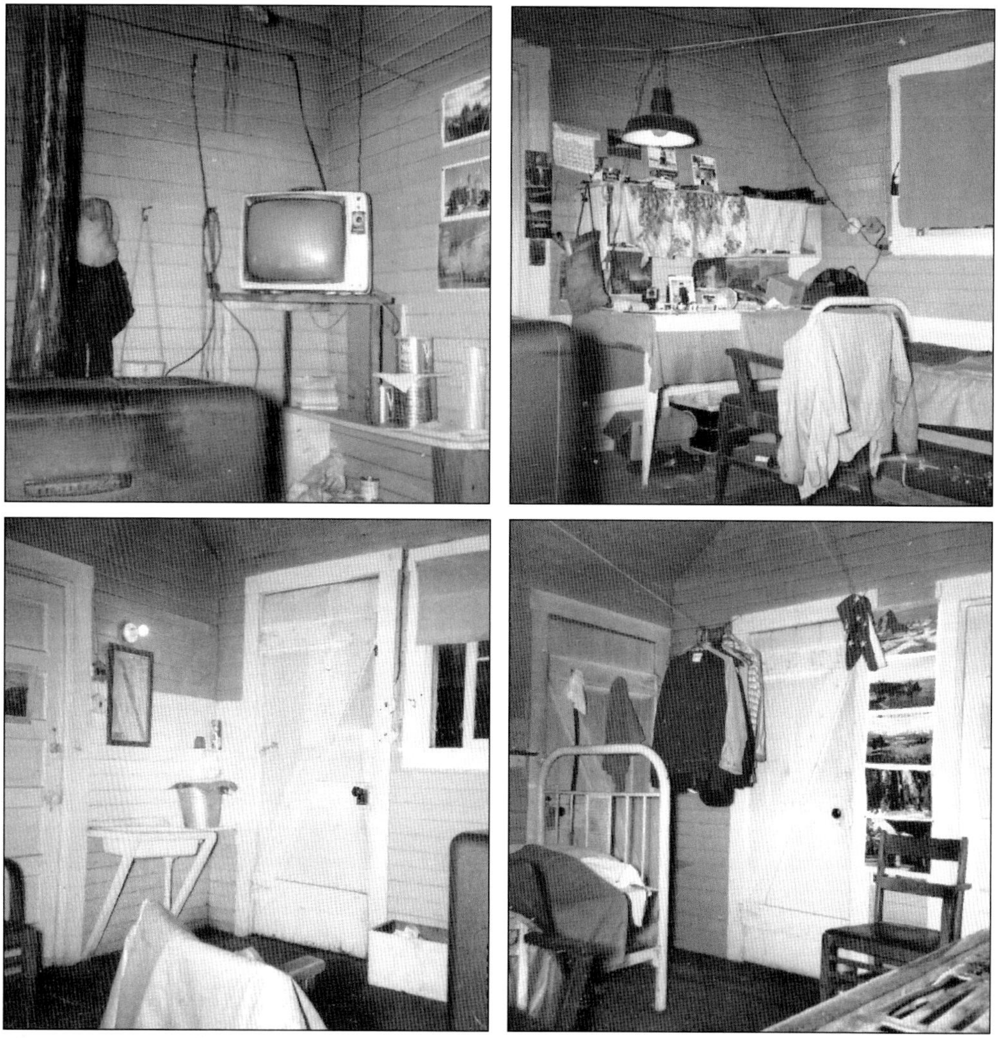

The snug, congenial, tolerant Camp 14 lifestyle suited Dale Hill just fine. His bunk area was located in one of the barn-red sidecar bunkhouses that lined the railroad tracks running through camp. These photographs offer a rare look inside the quarters of ordinary loggers. (All, DHC.)

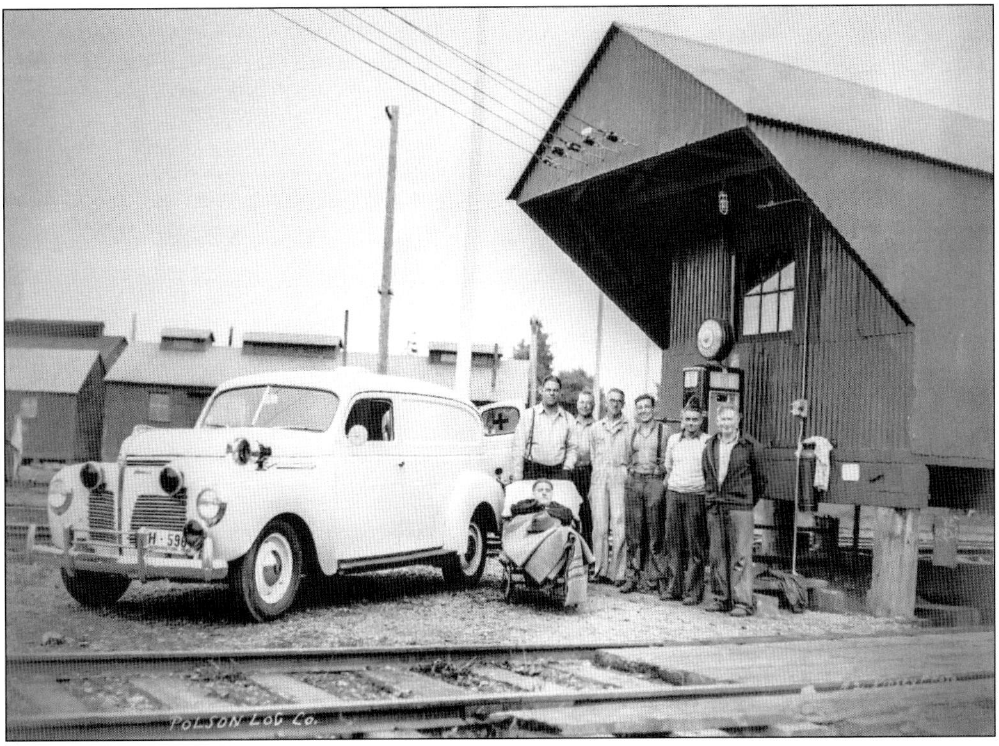

Polson Railroad camp was proud of its 1942 Plymouth aid car. The aid team members are, from left to right, Ralph Endreson, Les Rowe Sr., Ray Endreson, Jack Thompson, "Nig" (name unknown), and Harry Munsen. The "patient" is "Shorty" Groseclose. Most camps were at least an hour away from any medical facility. (PM.)

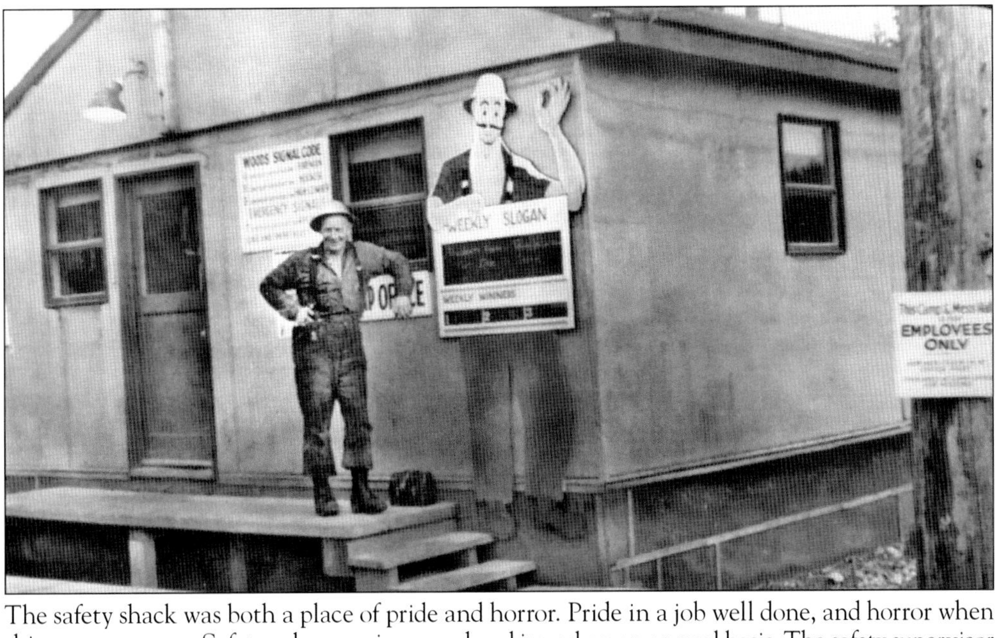

The safety shack was both a place of pride and horror. Pride in a job well done, and horror when things went wrong. Safety rules were improved and issued on an annual basis. The safety supervisor was capable of chewing out loggers over infractions in highly colorful language. (CHSGWC.)

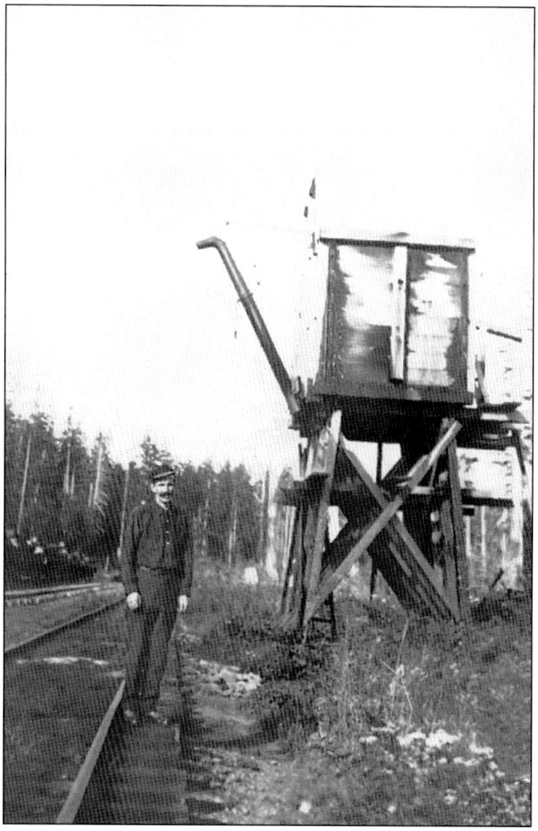

William Boeing, later known for his Boeing Airplane Company, began the Greenwood Logging Company in 1903 at the age of 22 to log east of Hoquiam. Here, a proud train crewman poses by the water tower. Greenwood was the first to use a gas railroad car (a 6-60) in its rail construction. (PM.)

Frank D. Hobi Logging Company's successful bid on the Quinault Reservation Cook Creek sale in 1922 necessitated a railroad camp location to house 250–300 men. Hobi was involved in interests ranging from "milking" pitch for commercial uses, to the Hobi Airways Flying Service on the port dock in Aberdeen. (MNB.)

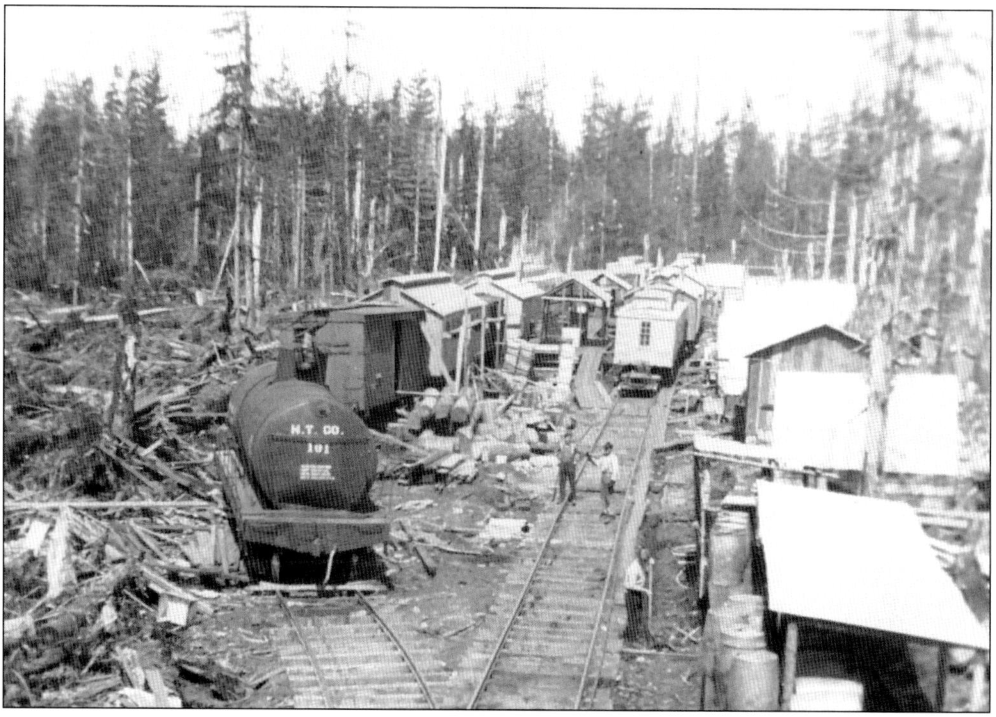

Five

STEAM TRAINS AND RAILS

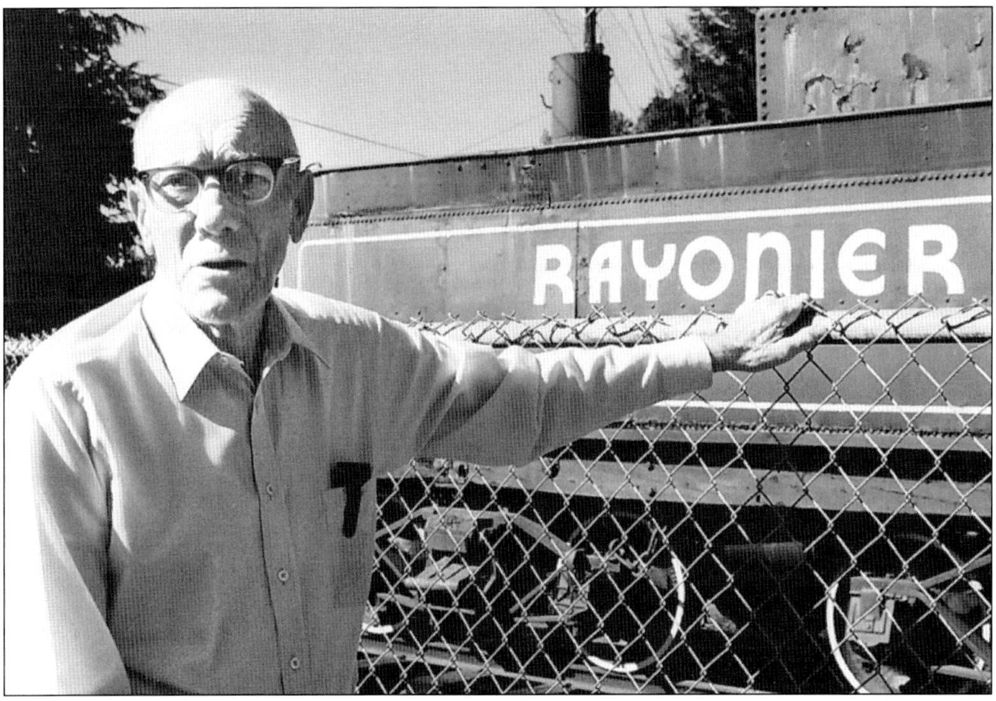

A railroad legend, engineer Thomas "Buster" Corrigan operated all 14 Polson engines out of Railroad Camp. He began at the Brooklyn Saginaw Camp, where he moved the longest train ever down the main line—88 loads. He drove a fully loaded string of log cars that had jumped the Polson track at Cook Creek, over a swamp and back on the tracks. (Gene Woodwick.)

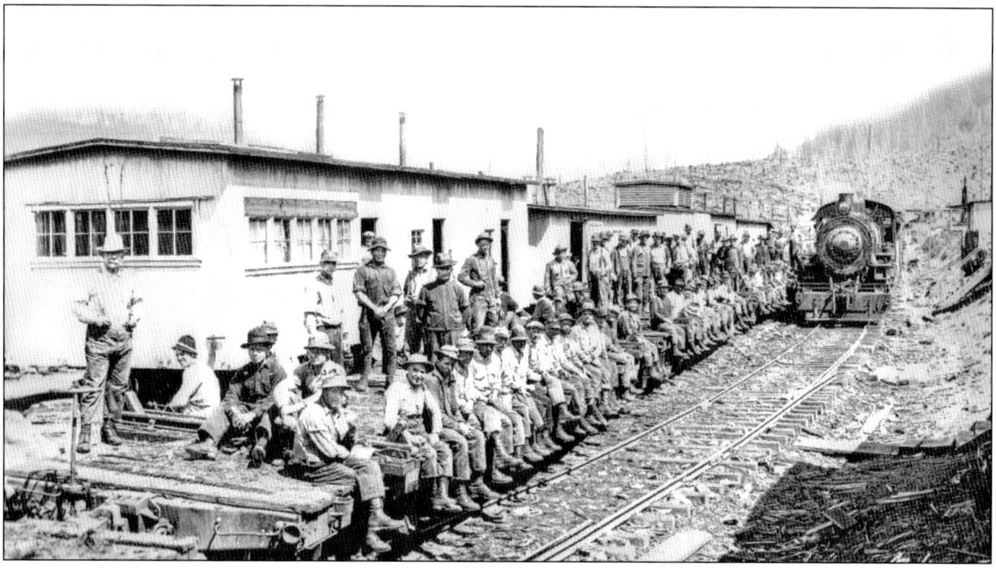

Steam trains vastly increased production in the woods and swelled camp populations. Here, the entire crew of Polson Camp 1 proudly poses for Clark Kinsey around 1910. The crew's steam engine stands in front of the camp housing. Polson Brothers Logging Company, begun in 1895 at Hoquiam, changed its name in 1903 to Polson Logging Company. It was bought out by Rayonier in 1948. (CHSGWC.)

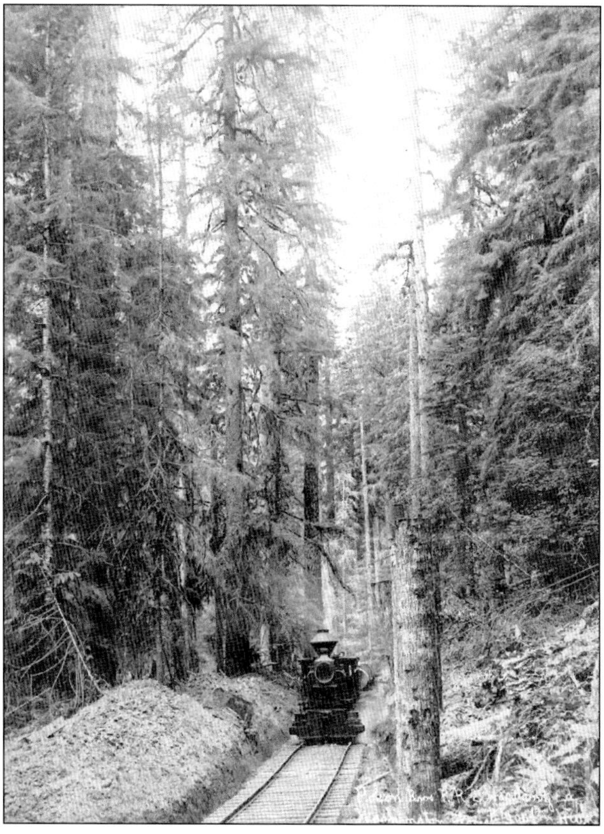

The famed Minnetonka steam engine was shipped around the Horn from Duluth, Wisconsin, to Kalamah, Washington, in 1871 to begin construction of the Northern Pacific Railway. The engine was sold to Polson and began work at New London and at the new railroad camp at Axford Prairie. Later, she was better known as "Old Farting Betsy." (CHSGWC.)

Chapin Collins (left), Montesano's *Vidette* newspaper editor, and Charles Teagle, former lokey engineer of the McCleary Lumber Company, exchanged viewpoints on "The Dinky," one of East County's first lokeys. It is now on display at the McCleary Beerbower Park. On December 31, 1941, the Simpson Timber Company of Shelton purchased the entire mill town. (CHSGWC.)

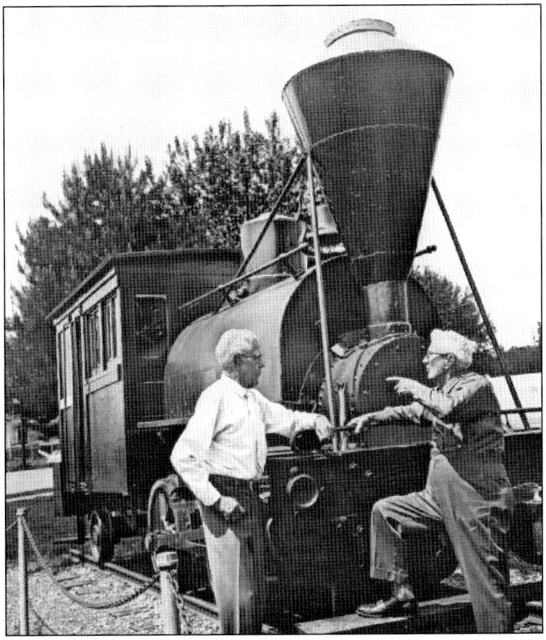

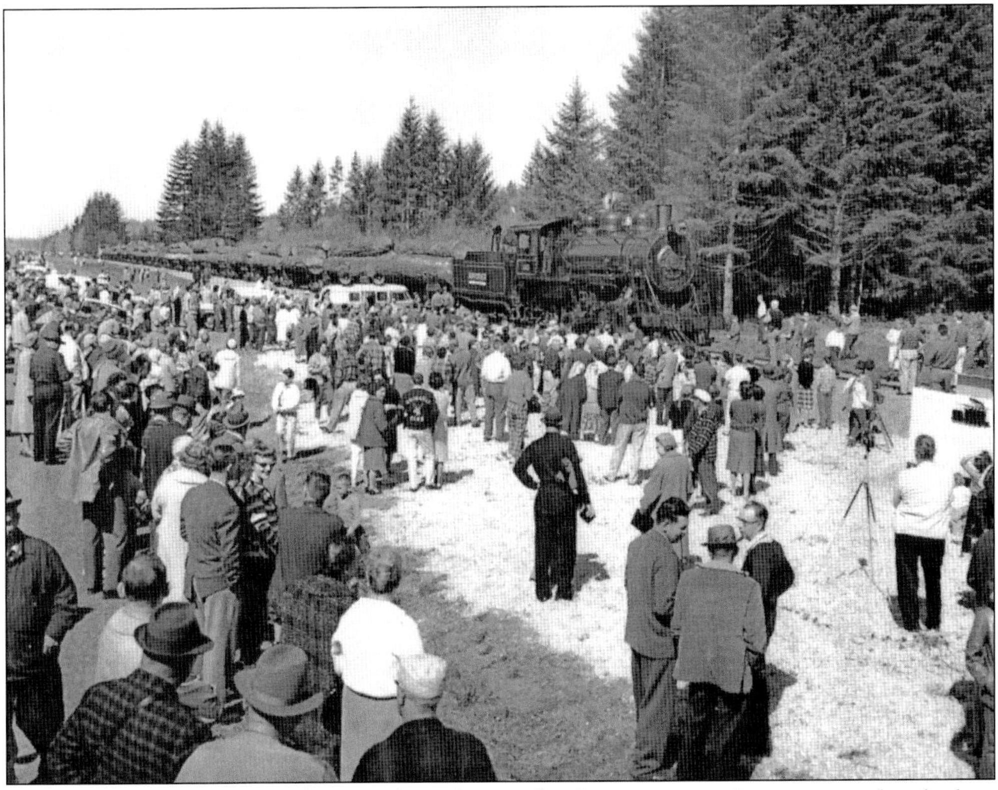

The Farewell to Steam Celebration drew thousands of spectators wishing to get a last look at their favorite steamers, some of which had spent 50 years working in the harbor's timber. No. 38 pulled the last load of logs out of the Humptulips country for the event. Old "8 Spot" was saved from the boneyard to serve as a reserve engine. (AMH.)

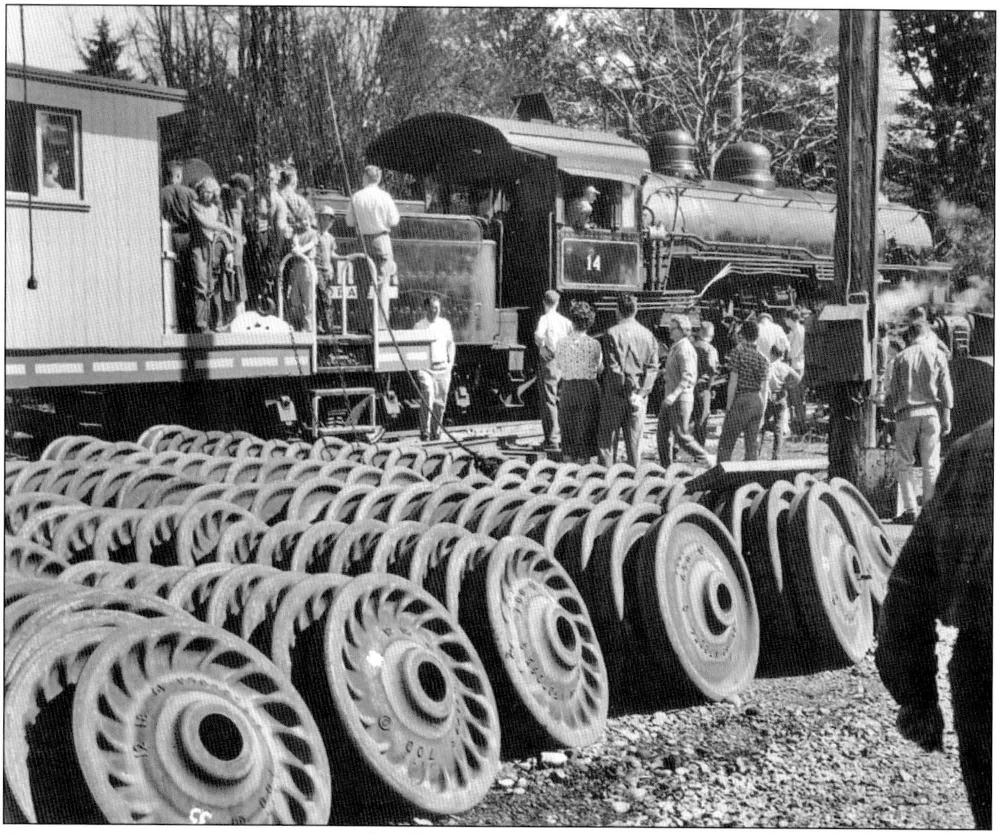

Rayonier's railroad camp overseer, J.K. "Bun" Lewis, was the personnel manager for Rayonier's Northwest Timber division, headquartered at Sappho. Lewis supervised the company's conversion from steam to diesel. During the conversion, Archie Inman was in the maintenance shop, supervised by foreman Ralph Andresen, with Adrian Julius as lokey mechanic. (AMH.)

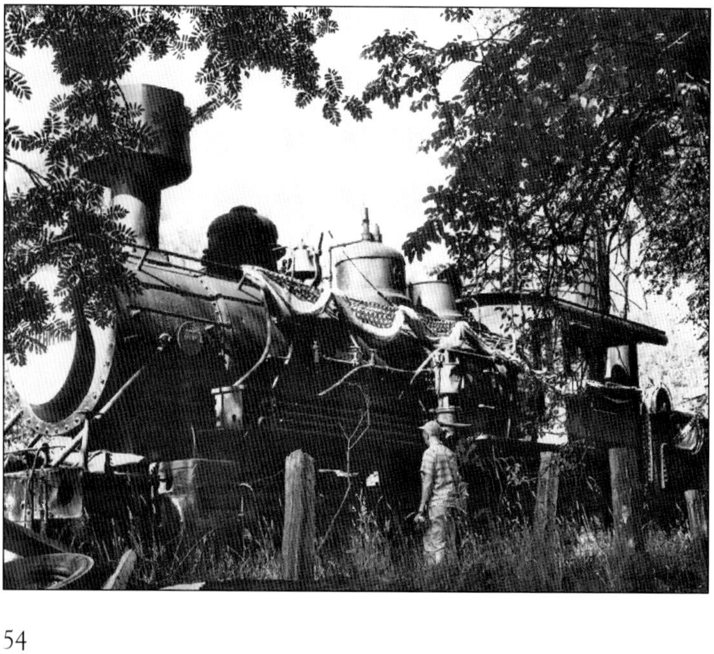

This Simpson Company locomotive was all gussied up for a 1956 local celebration. The earlier Simpson trains included the company's second lokey, a Climax named "Tollie," after Mrs. Sol Simpson. All of Simpson's locomotive engines were named for the towns they served and for mill personalities. (AMH.)

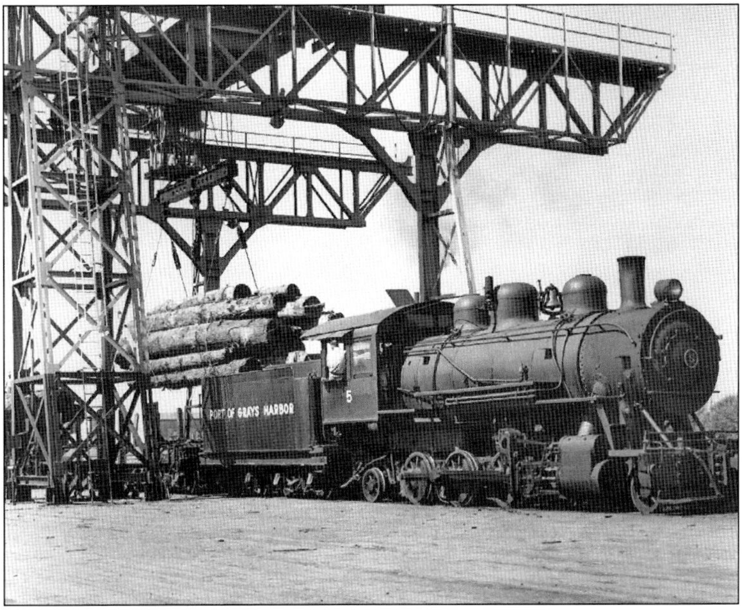

The Port of Grays Harbor's steam engine No. 5 is being loaded with logs in this Del Mulkey photograph. The Five Spot Porter 2-8-2 was built in 1924 for the Schafer Logging Company. It was later sold to the Mt. Rainier Scenic Railroad at Elby, Washington. (AMH.)

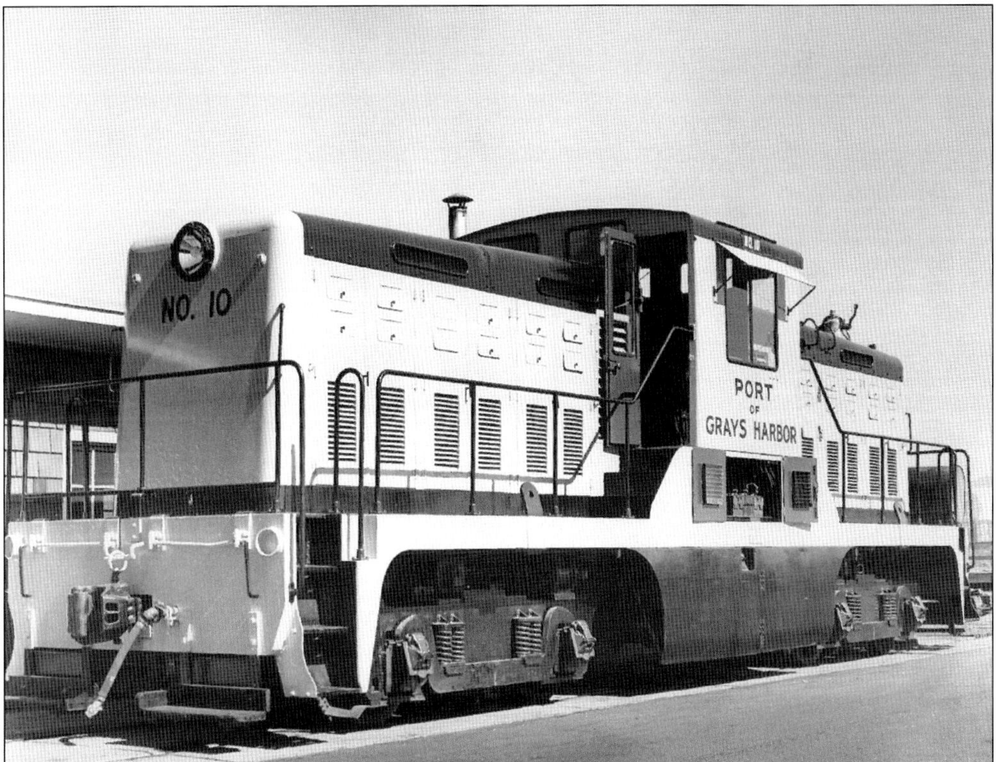

Rail has always been a key piece of infrastructure linking Grays Harbor's marine terminals to the producers of North America. In 1967, the port changed from steam to diesel power, exchanging the old No. 5 steamer to a No. 10 diesel. While it was praised for its economy, it never gained the affection local folks had for old No. 5. (AMH.)

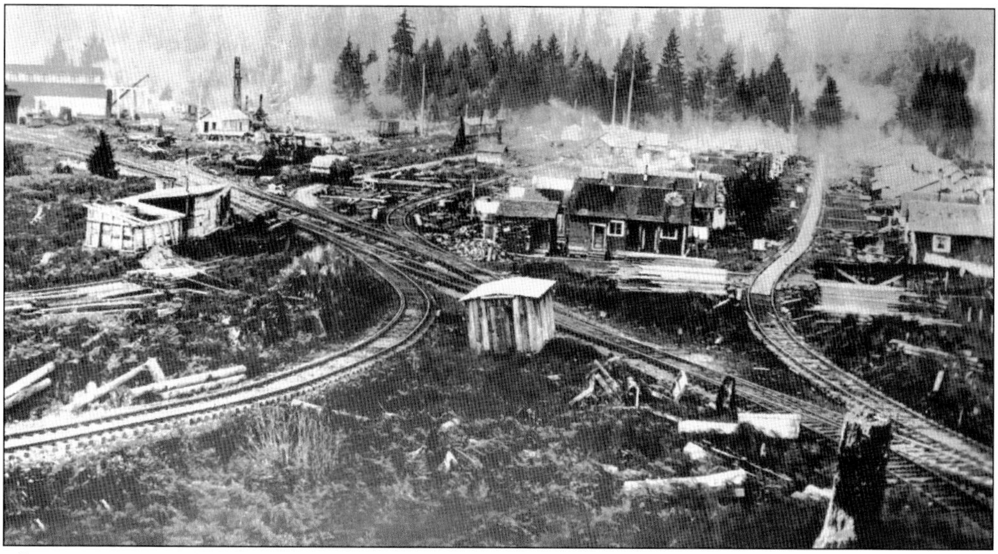

The Saginaw Timber Company, organized in 1909 by A.J. Morley and W.G. Hopkins, operated a 40-mile logging railroad. By 1934, the Saginaw Timber Company's Saginaw Logging Company operated at Brooklyn. On February 14, 1947, the company dissolved. It is likely that Buster Corrigan, who began his career as engineer, was at the throttle of the train barely seen at the top right. (AMH.)

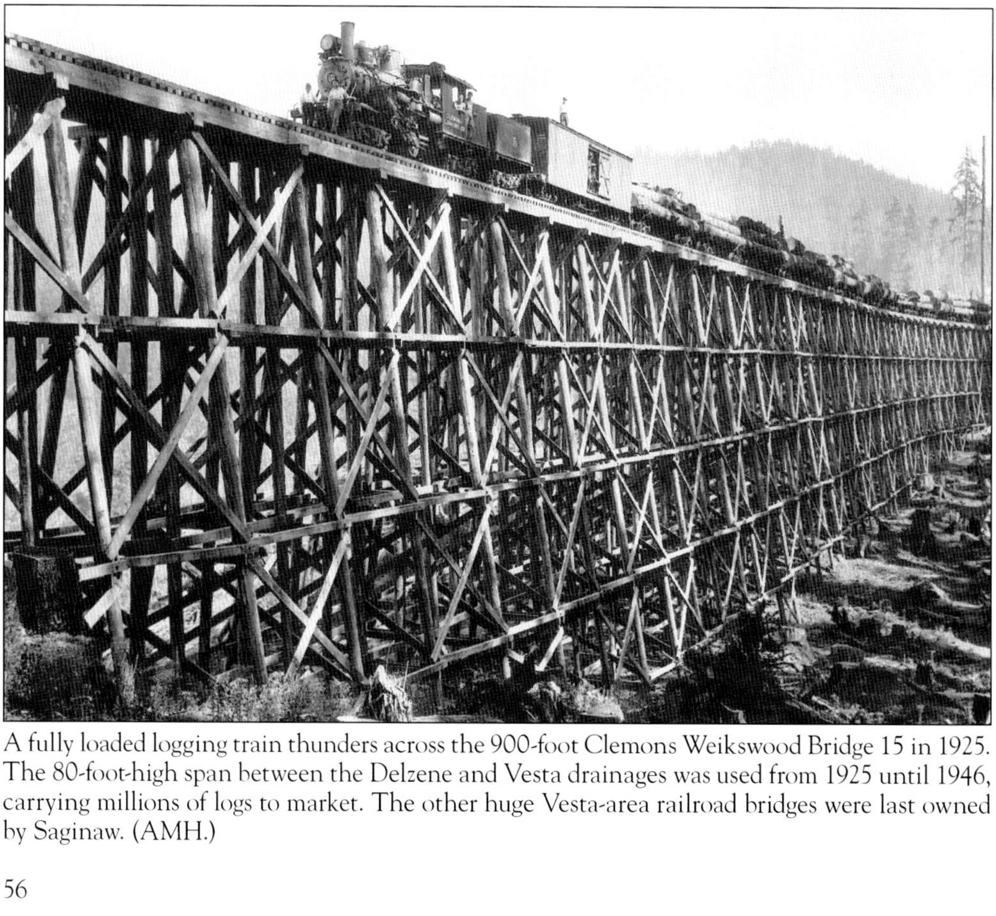

A fully loaded logging train thunders across the 900-foot Clemons Weikswood Bridge 15 in 1925. The 80-foot-high span between the Delzene and Vesta drainages was used from 1925 until 1946, carrying millions of logs to market. The other huge Vesta-area railroad bridges were last owned by Saginaw. (AMH.)

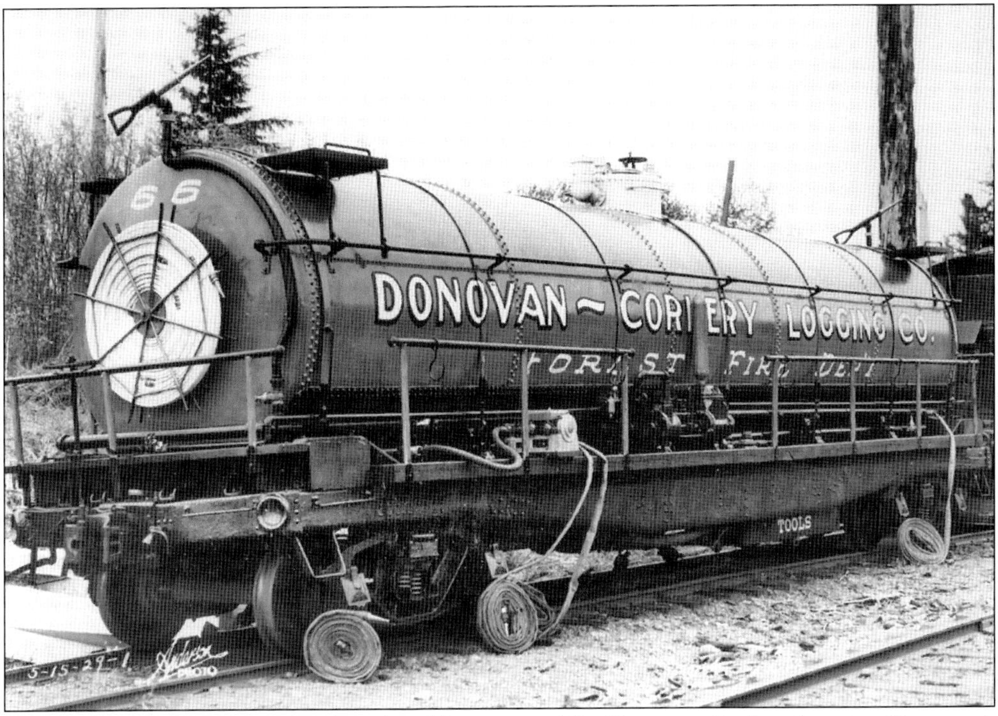

This is an unusual photograph of a railroad fire engine. The Donovan-Corkery Logging Company's forest fire department's engine is seen here on May 15, 1929. It was invaluable in putting out fires begun from steam-engine sparks or from other equipment uses. The company was known for its 2-8-2 Mikado engines. (AMH.)

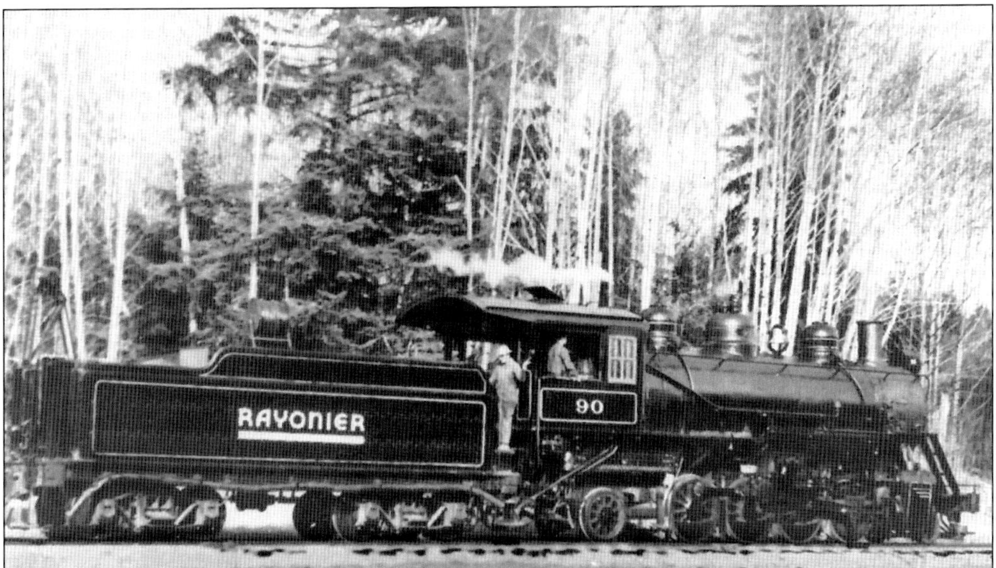

Rayonier Engine No. 90, with engineer Bobbie Rogers, another famed Rayonier lokey driver, at the throttle, is seen near Humptulips. Charlie Lane was firing, Fred Nix served as brakeman, and Charlie Petrovich was the second brakeman. The steam engine and its crew were familiar to families in the North County area. (AMH.)

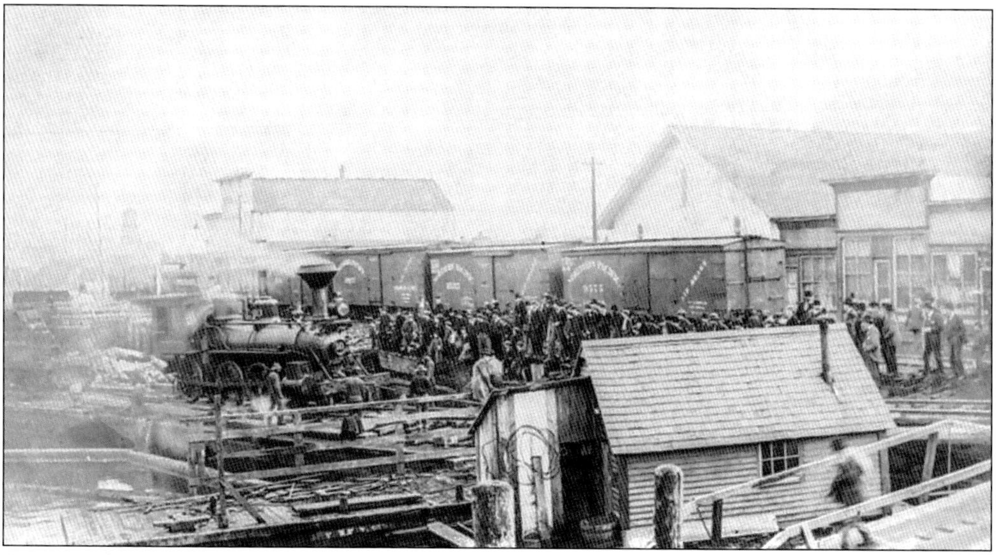

A new Polson locomotive is unloaded from a scow at the company's Hoquiam Eighth Street landing. Scowing heavy equipment up both the Little and Big Hoquiam Rivers was not without risk. Hidden logs in the water could dump equipment into the brink. (AMH.)

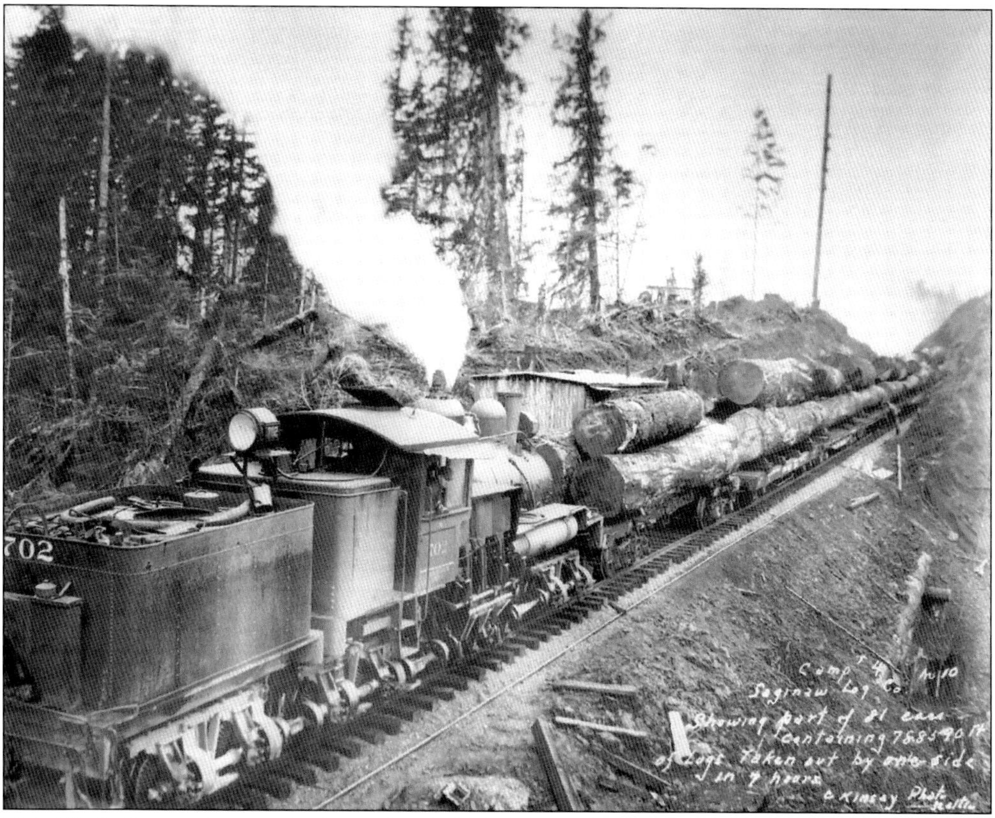

This Clark Kinsey photograph shows the 7,684,690 feet of logs that filled 21 cars. They were taken out of the Saginaw Camp 10 side in just nine hours. This was an amazing feat, considering it took place at the dawn of the era of power saws and other modern methods. (AMH.)

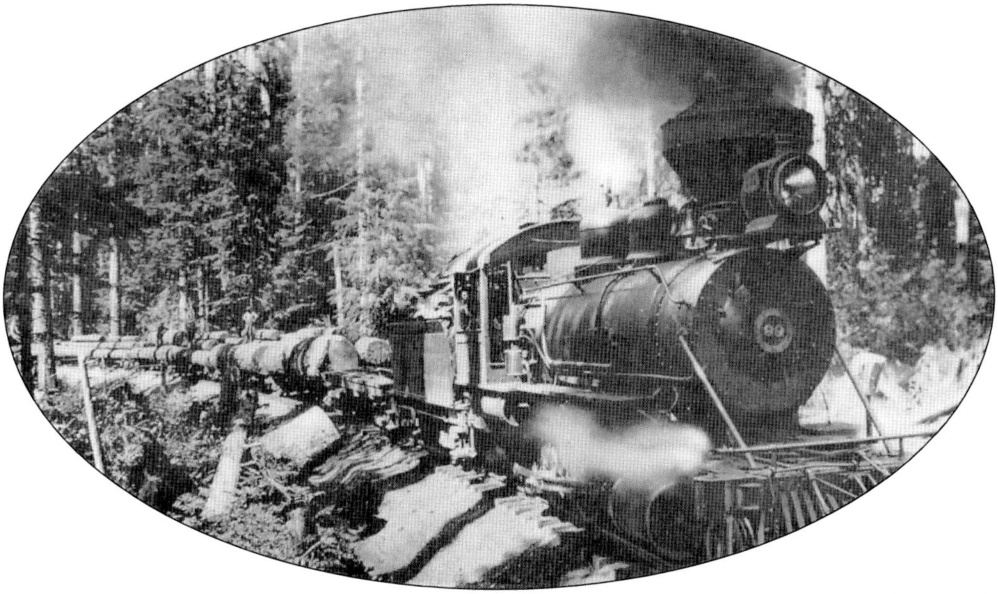

One of the first Polson trains, with Alex Polson at the throttle, is shown in this pre-1900 photograph postcard. Alex was the hands-on guy at the Polson Brothers shows. He was also not averse to using his fists in street fights in Hoquiam if he was challenged. (CHSHWC.)

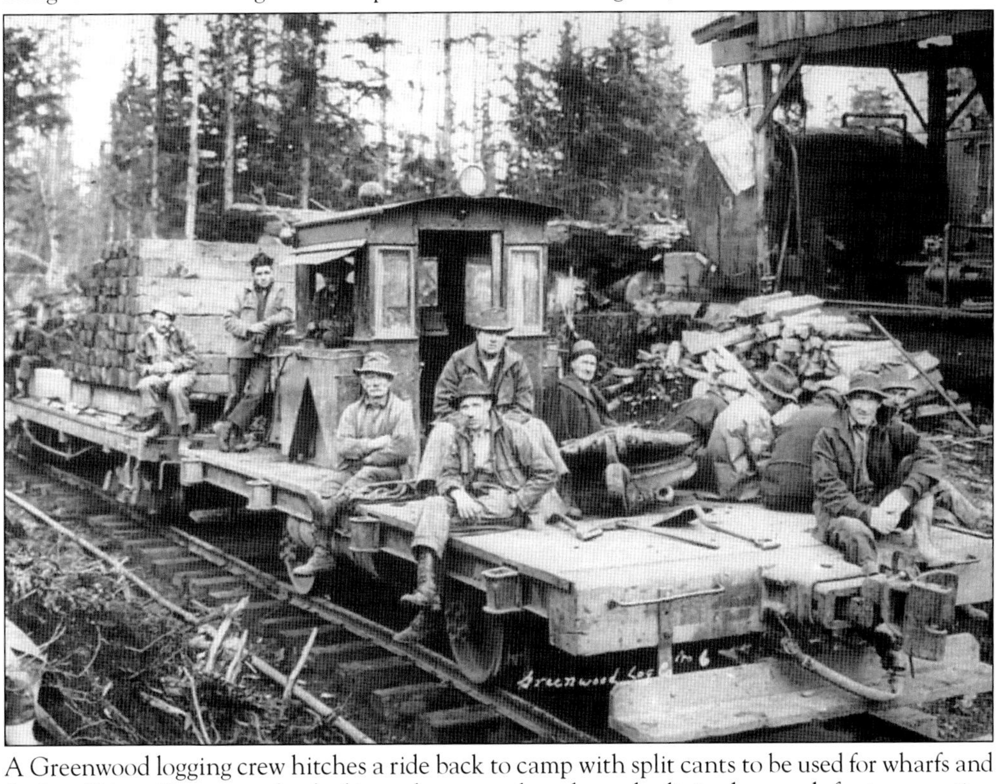

A Greenwood logging crew hitches a ride back to camp with split cants to be used for wharfs and deck cribbing. Cants were bucked into the proper lengths and split in the woods for transporting directly to harbor wharfs for shipping coastwise (the North and South American Pacific coastline) and to foreign ports. (PM.)

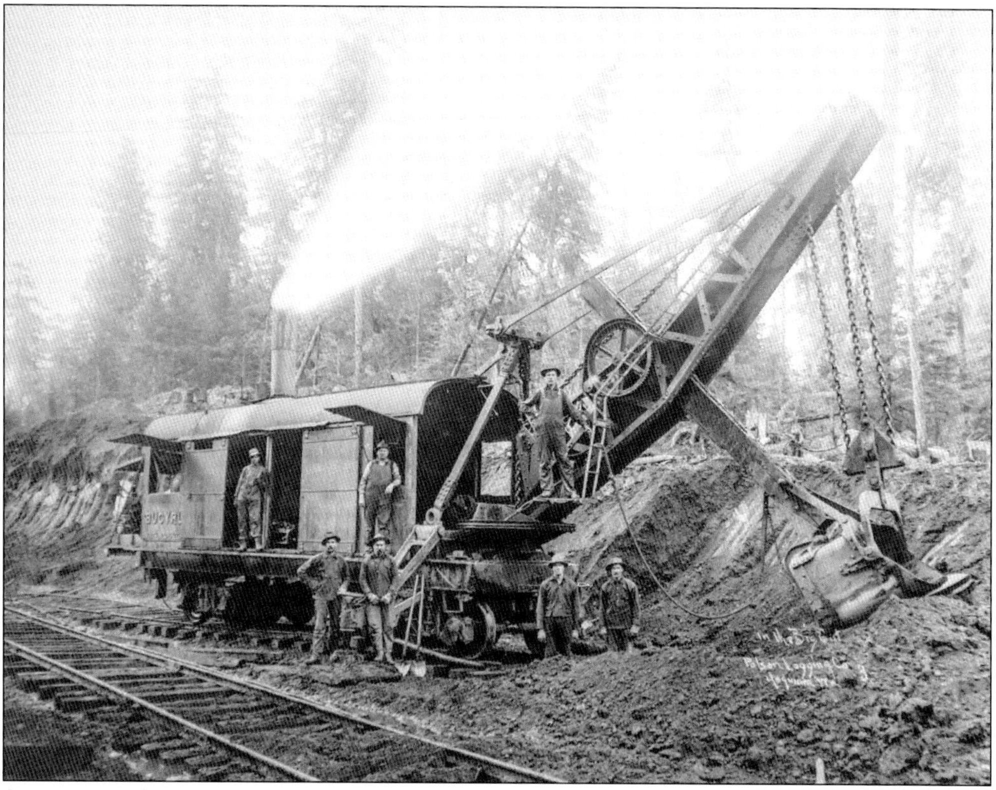

Steam pours from this rail-laying equipment of an unknown early-day harbor logging company. The increased use of steam trains for logging purposes called for greater amounts of trackage. This type of rail-laying equipment was used on all US rail lines until the 1950s. (PM.)

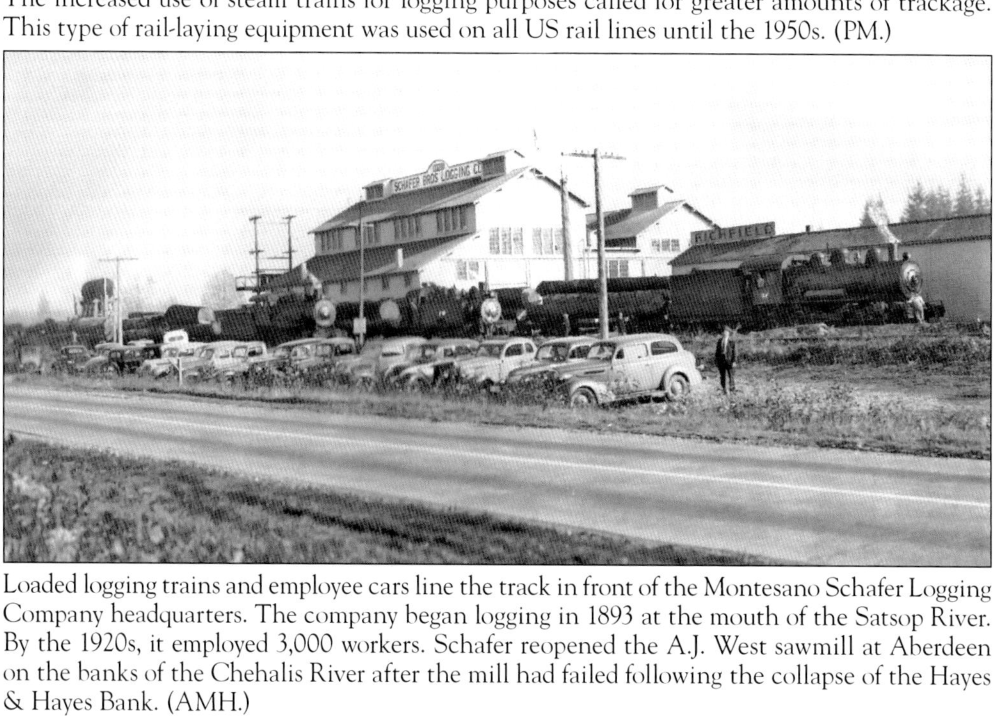

Loaded logging trains and employee cars line the track in front of the Montesano Schafer Logging Company headquarters. The company began logging in 1893 at the mouth of the Satsop River. By the 1920s, it employed 3,000 workers. Schafer reopened the A.J. West sawmill at Aberdeen on the banks of the Chehalis River after the mill had failed following the collapse of the Hayes & Hayes Bank. (AMH.)

Six

Gyppos and White Hats

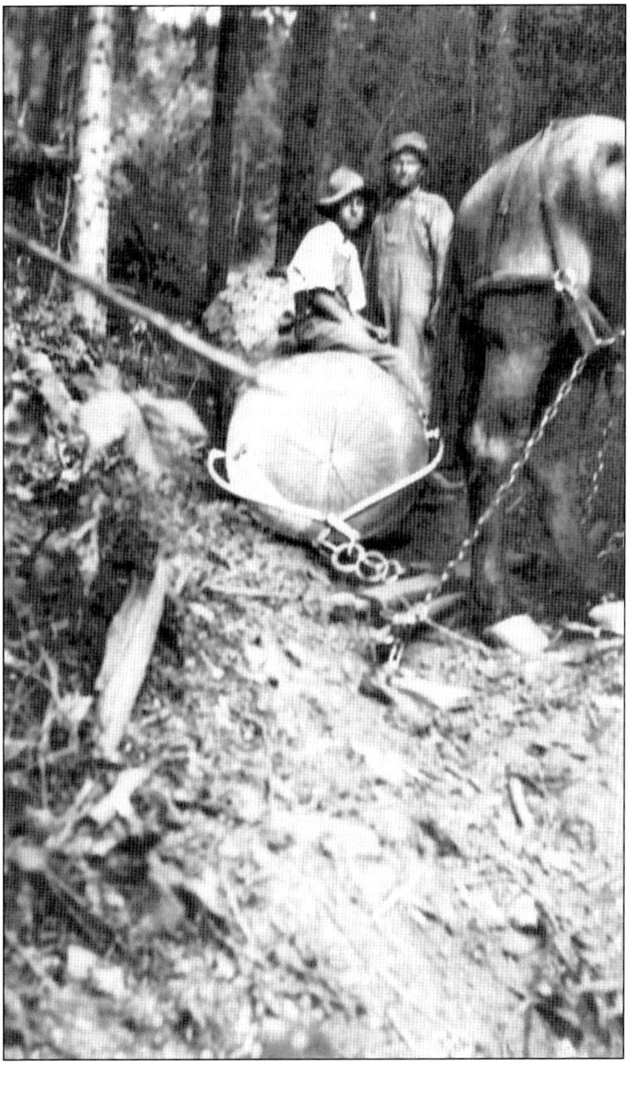

All it took for beginning horse logging were two guys, one horse, no wire rope, and just chains. The smaller size of logs was dictated by mill pricing that scaled logs at the small, top end. The mills did not pay for anything over 58 inches in diameter—a pretty good deal for mills, but not for logging companies. (CHSGWC.)

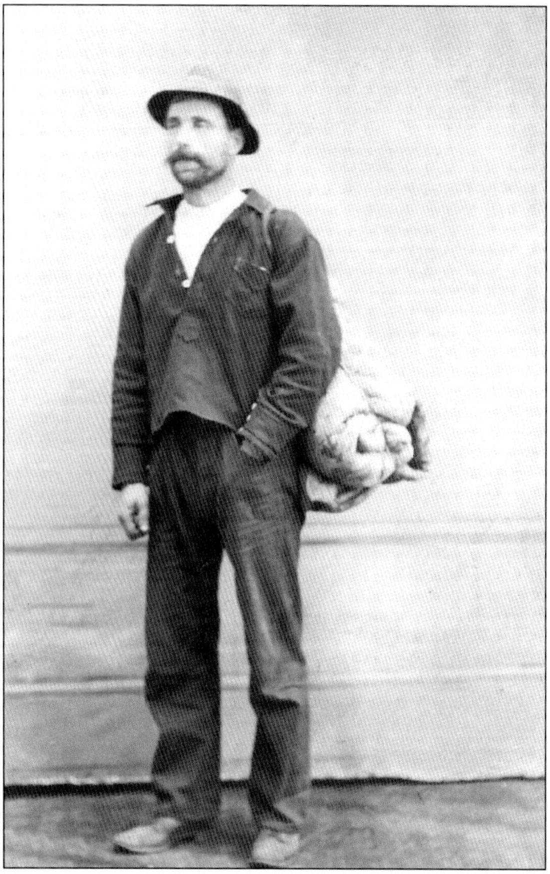

The "man catcher" (employment office) must have caught this fellow. He posed with his bedroll for an Aberdeen street photographer before heading out for work in a new logging camp. The man catcher charged $1 for a job placement and advertised jobs on a sidewalk chalkboard. (CHSGWC.)

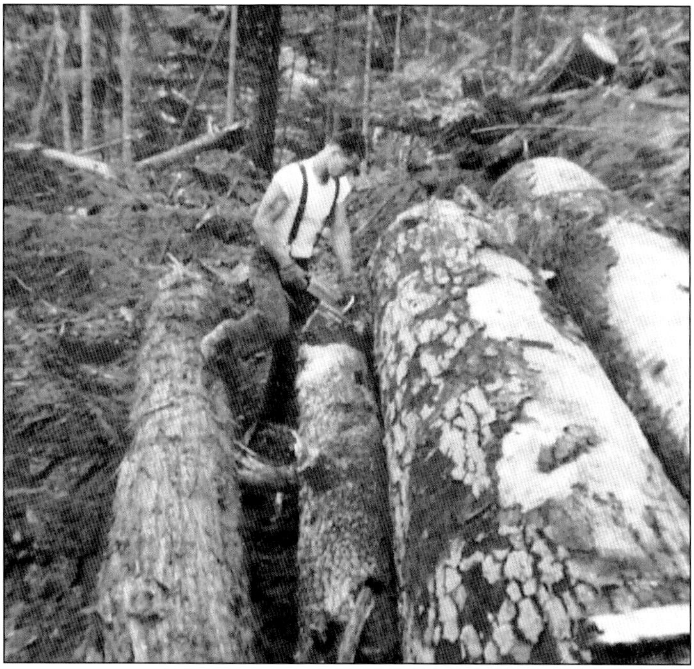

The woods of Grays Harbor were full of men like Rand Auckland, hired for their particular ability in logging jobs. The Olympic Peninsula was one giant community of men and gyppo (freelance) operators, hired to use their specialized skills at a multiplicity of jobs. (Gene Woodwick.)

This gyppo checks out the salvage value of blowdown timber from the 1962 Columbus Day storm. That storm was exceeded in destructiveness by the one on December 3, 2007, that slammed the harbor with wind gusts of 80 to 100 miles per hour. Weyerhaeuser (Weyco) timberlands were hit hard, with hundreds of acres of forestlands flattened. (Gene Woodwick.)

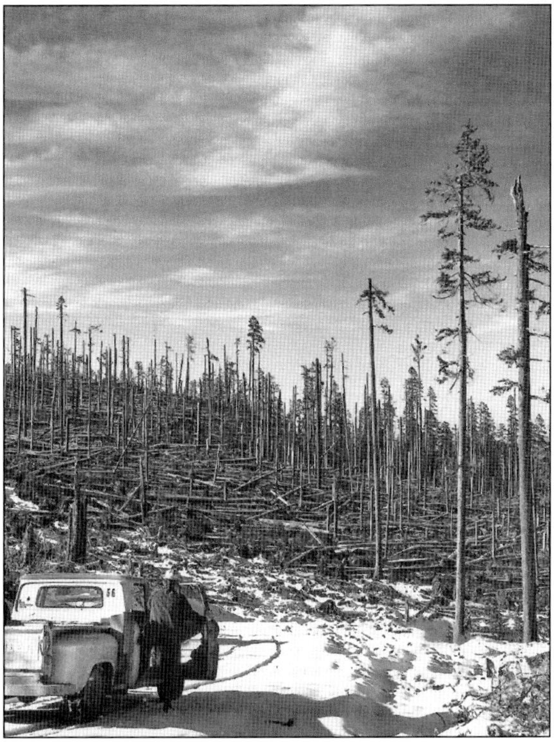

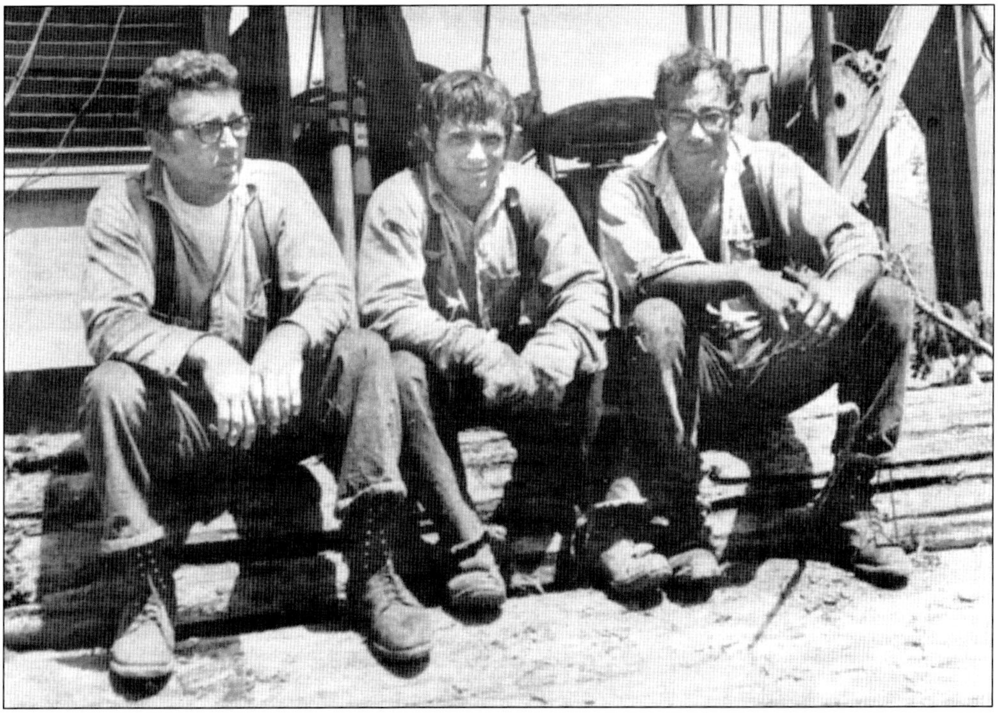

This 1960s *Elma Chronicle* newspaper photograph shows gyppo loggers from C&R Logging Company. Fred Rakevich (left), Rick Rakevich (center), and Ed Comfort take a break from contract logging on Swede Hill above Porter and Malone in East County, on Weyerhaeuser Company lands.

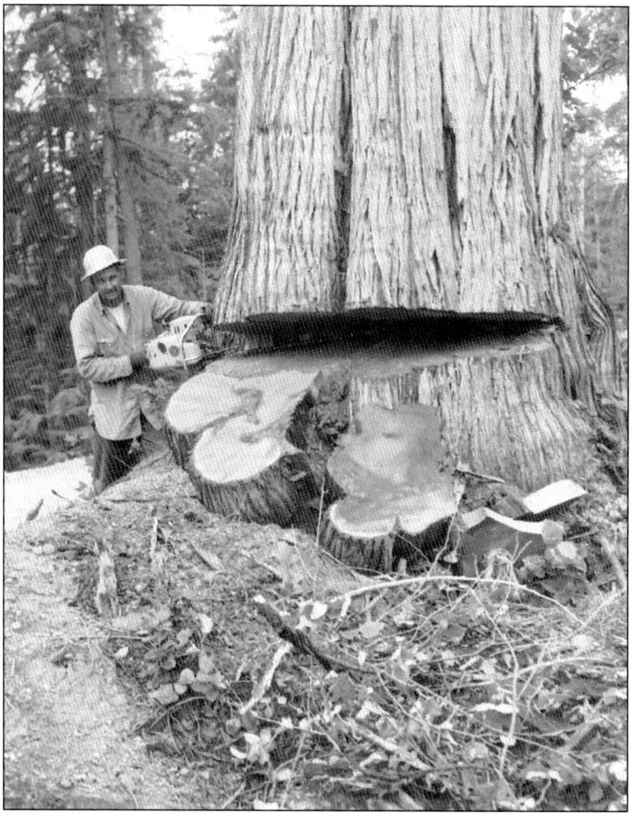

Leland Mann takes on this 10-foot diameter cedar in North County with his long bar McCullough around 1970. A good faller was always in demand on the Olympic Peninsula. Most fallers were capable of sharpening their own saws because their safety and life depended upon good cutting equipment. (MNB.)

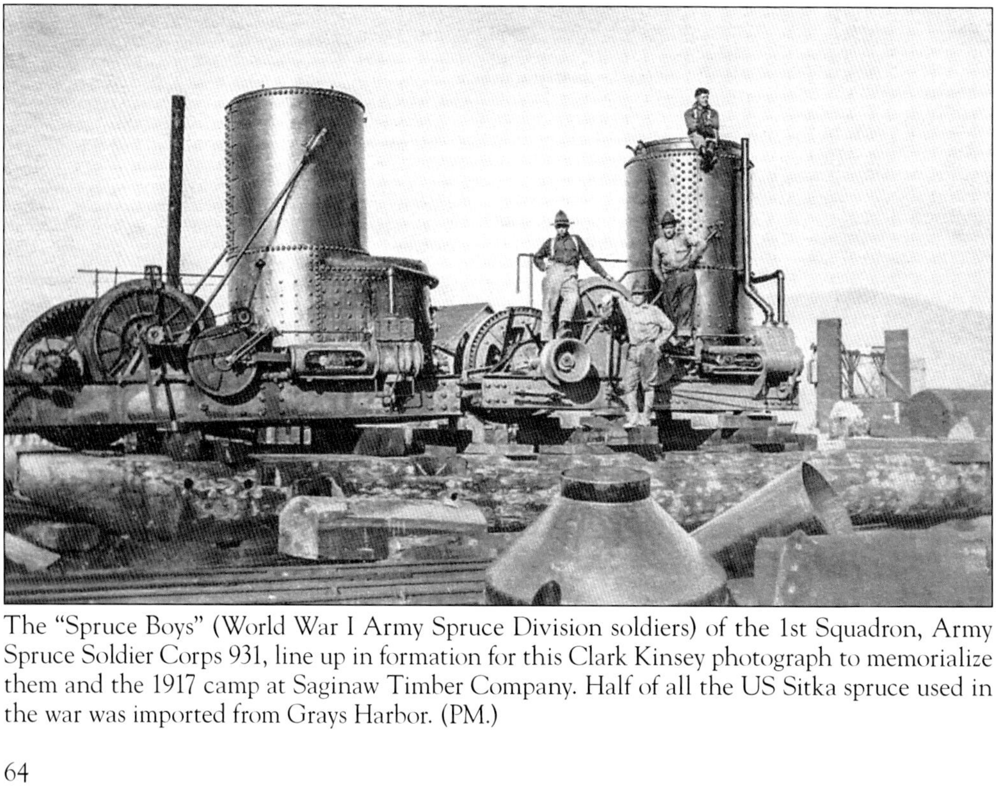

The "Spruce Boys" (World War I Army Spruce Division soldiers) of the 1st Squadron, Army Spruce Soldier Corps 931, line up in formation for this Clark Kinsey photograph to memorialize them and the 1917 camp at Saginaw Timber Company. Half of all the US Sitka spruce used in the war was imported from Grays Harbor. (PM.)

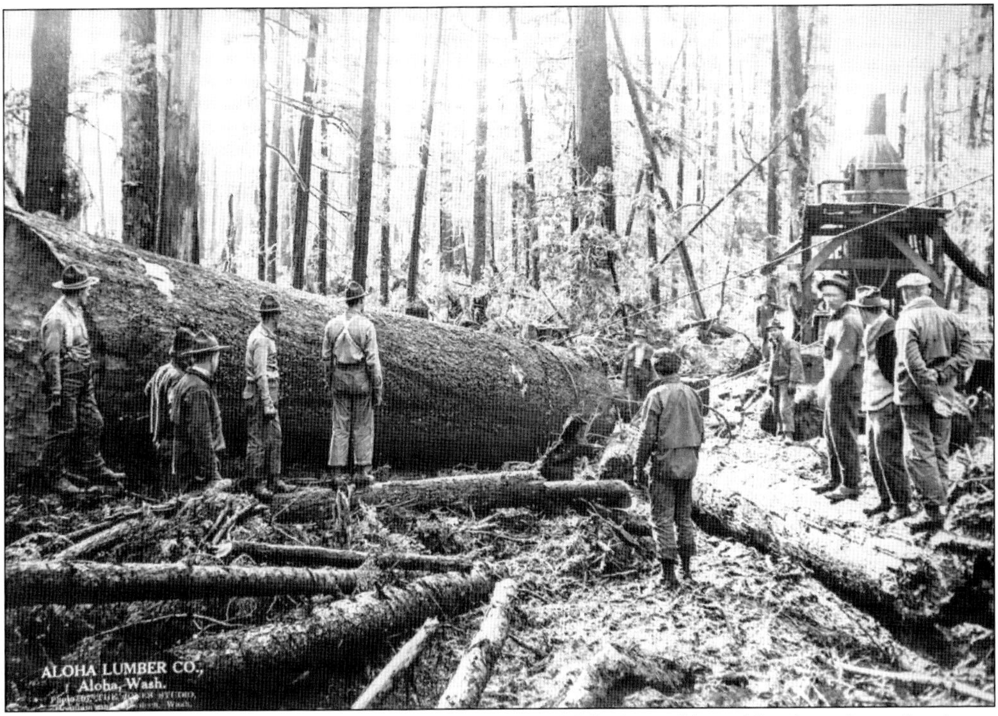

Springboards, misery whips (the so-called Swedish fiddles), and hand-bucking brought down spruce logs measuring eight feet in diameter with almost straight grain. They were split using 10 iron and 10 yew wedges. Splinters were cut off with "spuds" (chisels), then quartered. The center, heart wood cant was used for airplane production. (PM.)

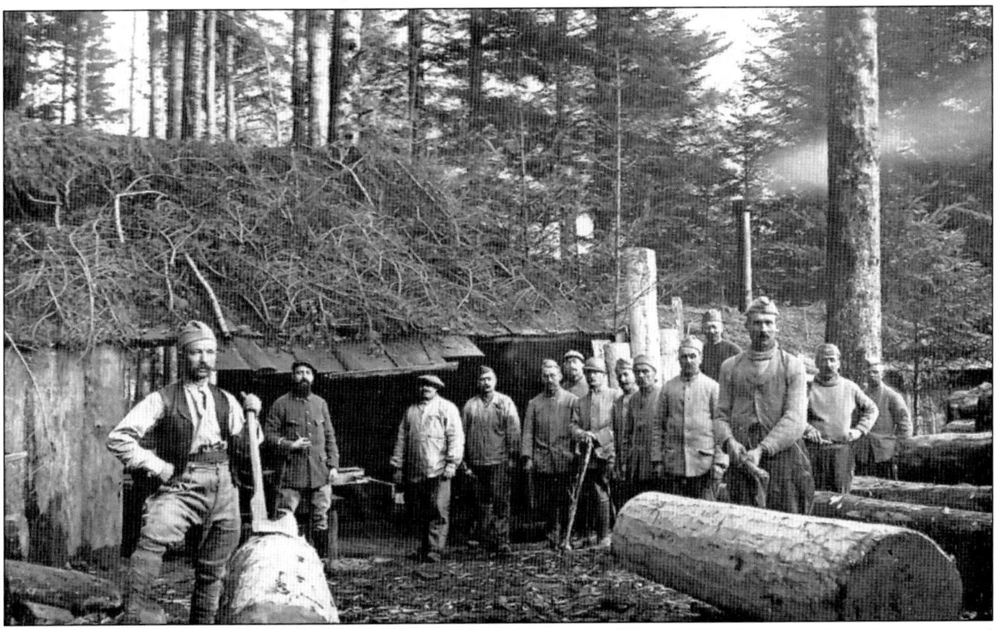

Local Grays Harbor soldier boys were shipped to France in World War I to put their logging and mill backgrounds to use in coordination with French troops. Their skills as mill workers and loggers was considered far more important than "fighting the Huns." (WSM.)

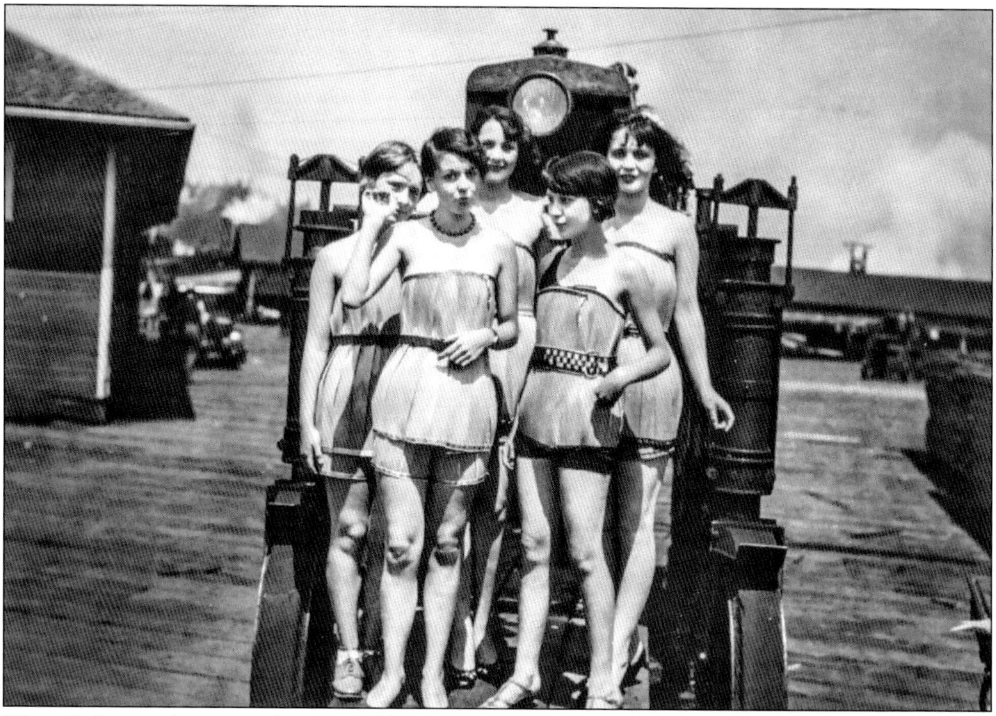

These ladies made national news in film and print media when they appeared in bathing suits of spruce veneer as a 1924 publicity stunt for the Grays Harbor Veneer Company's Hoquiam Wood Week. Willa Biggs, 16, and her future sister-in-law Julia Maring Garrison were among the first "Spruce Girls," so named after the World War I Spruce Boys. (PM.)

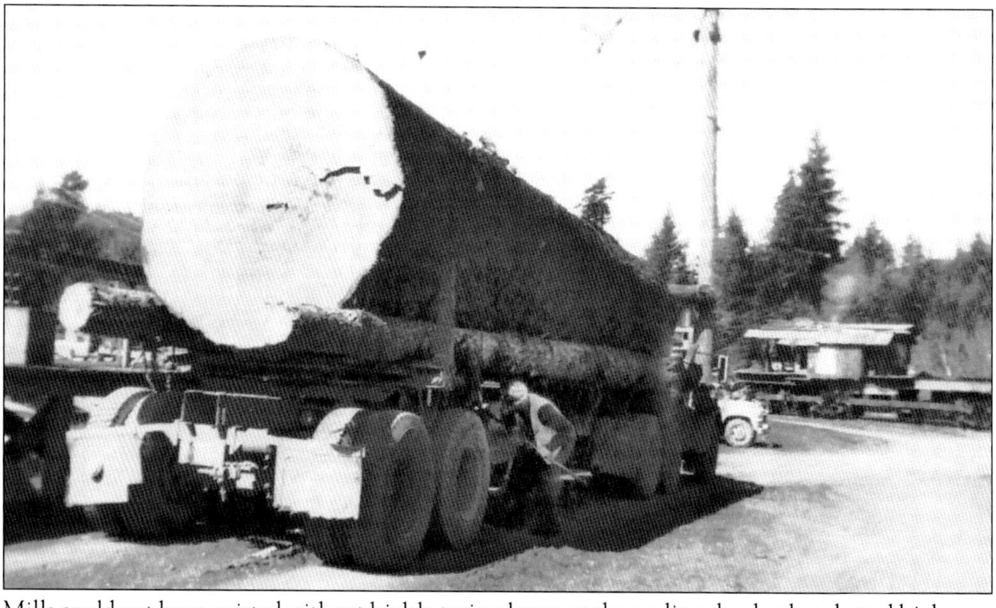

Mills could not have existed without highly trained men at the scaling shacks that dotted highways and mill yards. Grays Harbor Scaling was the fourth-largest such firm in the United States. Scalers, among the most respected timber workers, served as a neutral third party to determine quality and quantity of lumber. (Gene Woodwick.)

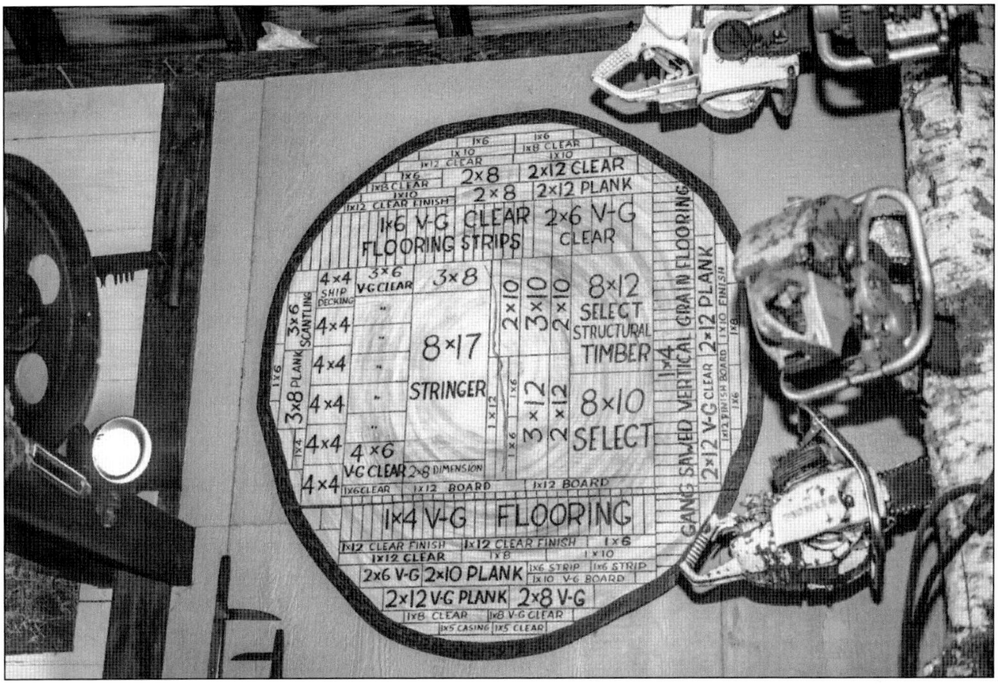

The power chainsaws seen here in a Willapa Seaport Museum display provide an indication of the size of logs entering the mill in the 1960–1970s. This log section illustrates the type and amount of lumber produced from such a log. (Brian Woodwick.)

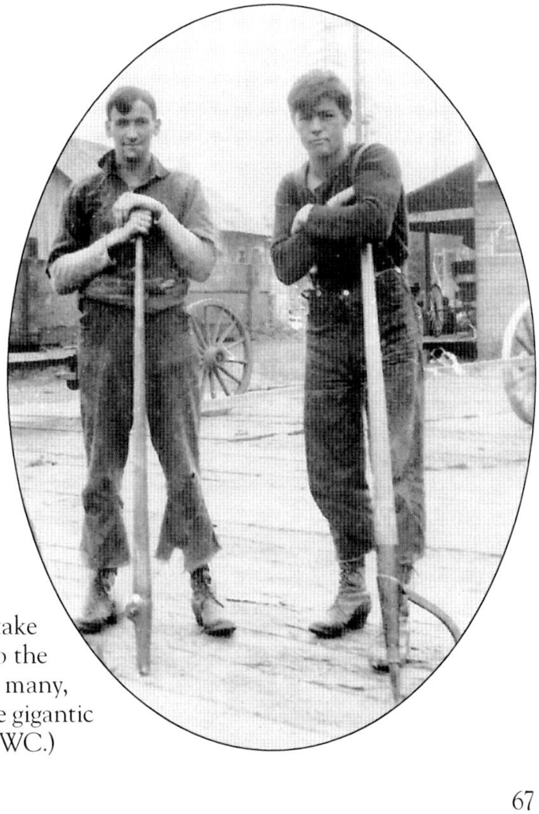

Peaveys in hand, these young mill workers take a break from manhandling logs coming into the mill. These log "wrestlers" knew just how many, and what size, logs to send up the ramp to the gigantic cutoff saw just inside the mill door. (CHSGWC.)

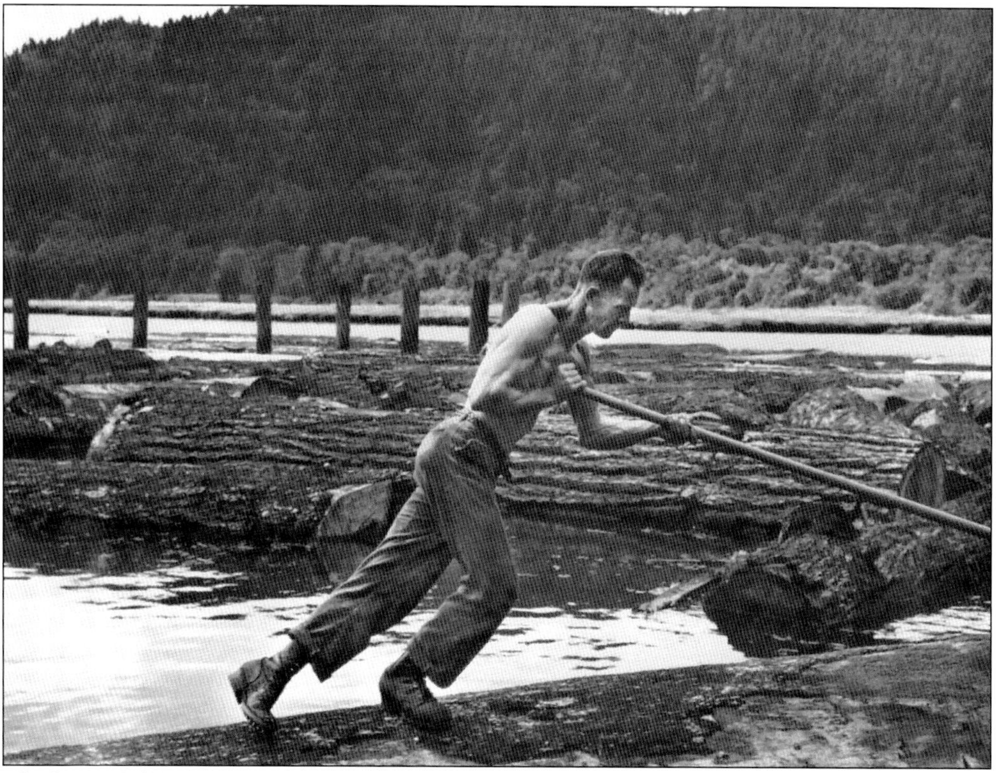

The hero of all logging kids was the graceful boom man, who pushed massive logs around rivers and millponds using a 20-foot-long pike pole with its spiked and hooked end. A boom man could turn the air blue if he caught a kid on the booms dangerously mimicking his job. (AMH.)

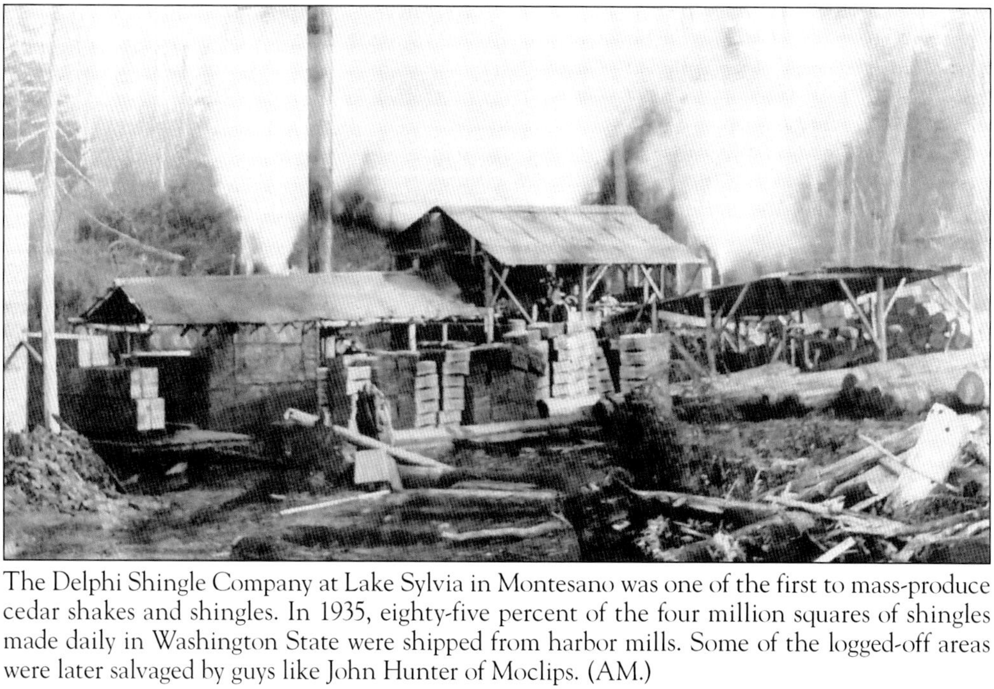

The Delphi Shingle Company at Lake Sylvia in Montesano was one of the first to mass-produce cedar shakes and shingles. In 1935, eighty-five percent of the four million squares of shingles made daily in Washington State were shipped from harbor mills. Some of the logged-off areas were later salvaged by guys like John Hunter of Moclips. (AM.)

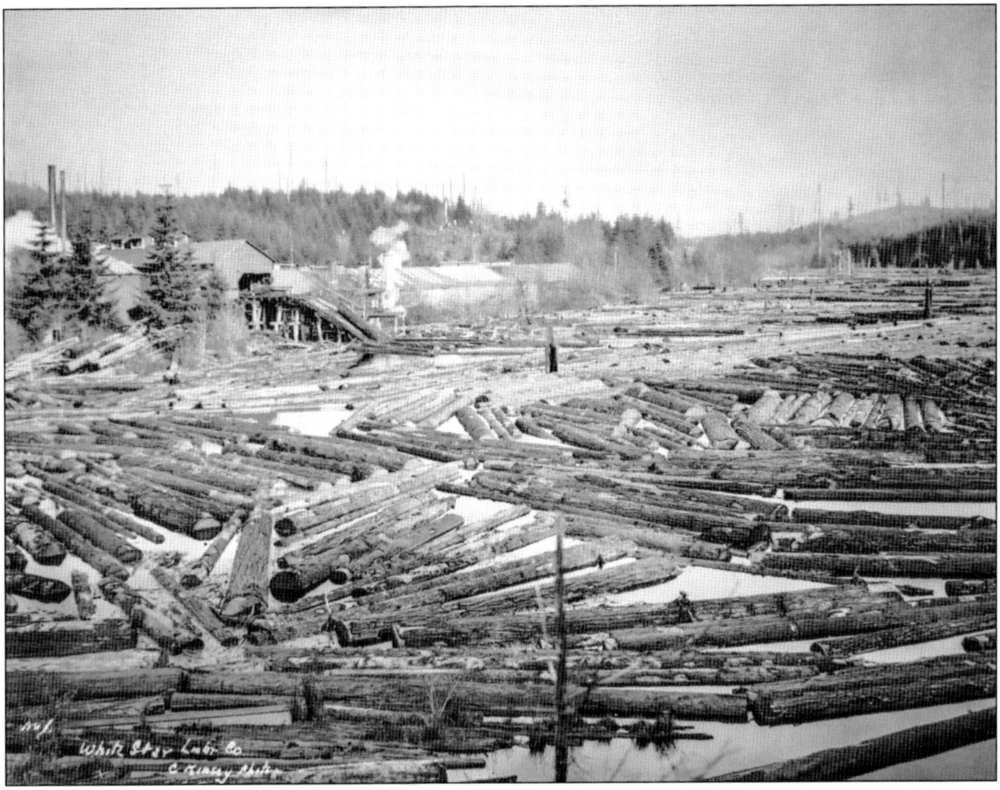

This millpond overflowing with logs provides a scale of production at the White Star Lumber Company, which began in 1909 east of Montesano. J.R O'Donnell was the manager when the company was turning out 80,000 feet of lumber and 100,000 shingles in 10-hour days. (PM.)

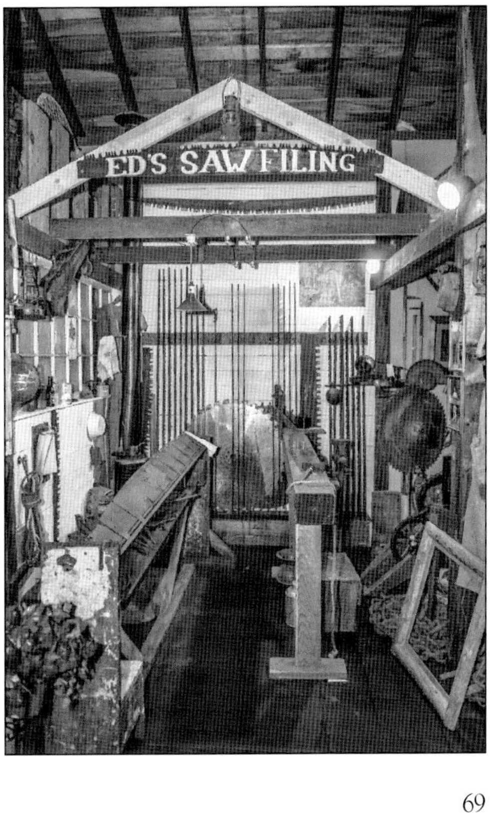

One of the most highly skilled and highly paid positions in any mill was that of the saw filer. A poorly sharpened saw not only slowed production but could maim or kill a worker. This Willapa Seaport Museum display is an accurately reconstructed saw filing shop, used by both Ed Janhusen Jr. and his father, Ed. Sr. (Brian Woodwick.)

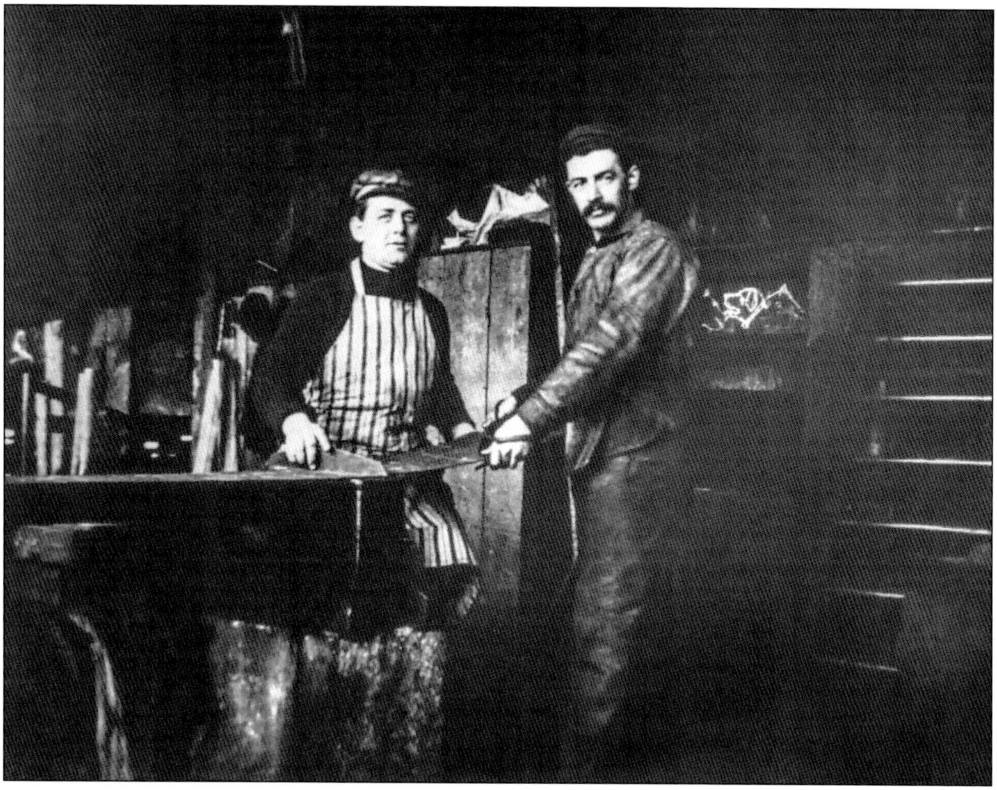

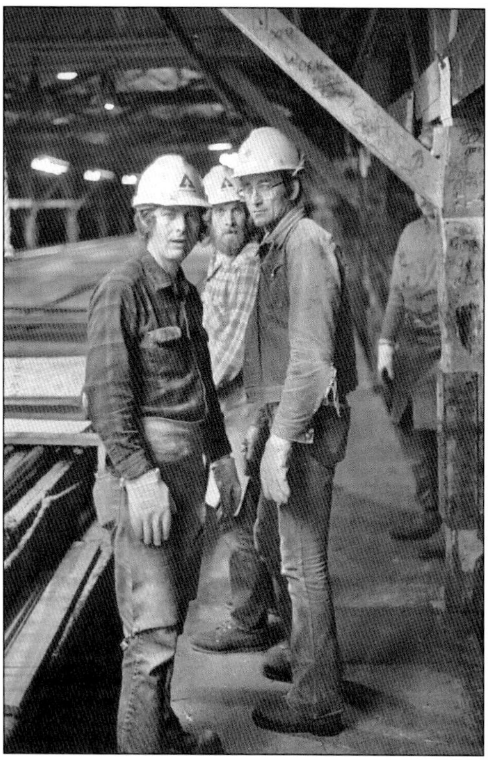

Piloman DeLorme (left) opened his saw shop in 1932 after working as head filer at Anderson & Middleton Mill. His sons, Fred A. and Roland, were head filers at Schafer Brothers and later for the Weyerhaeuser Company. (Larry and Marcia DeLorme.)

These workers, getting ready for a shift change at the Weyerhaeuser Mill, demonstrate the variety of ages of those who found employment in the milling industry. For many years, it was a beginning job with good pay, a satisfying occupation, and one with good benefits and retirement packages. (Brian Woodwick.)

This Weyco saw filer had his own lair for his careful and skilled work, so necessary for high production. The room's walls were a bulletin board for scatological humor and appreciation of pulchritude that predated today's political correctness. (Brian Woodwick.)

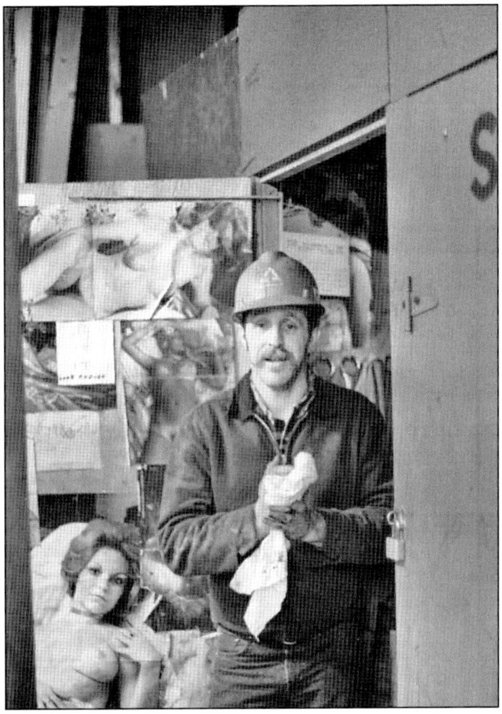

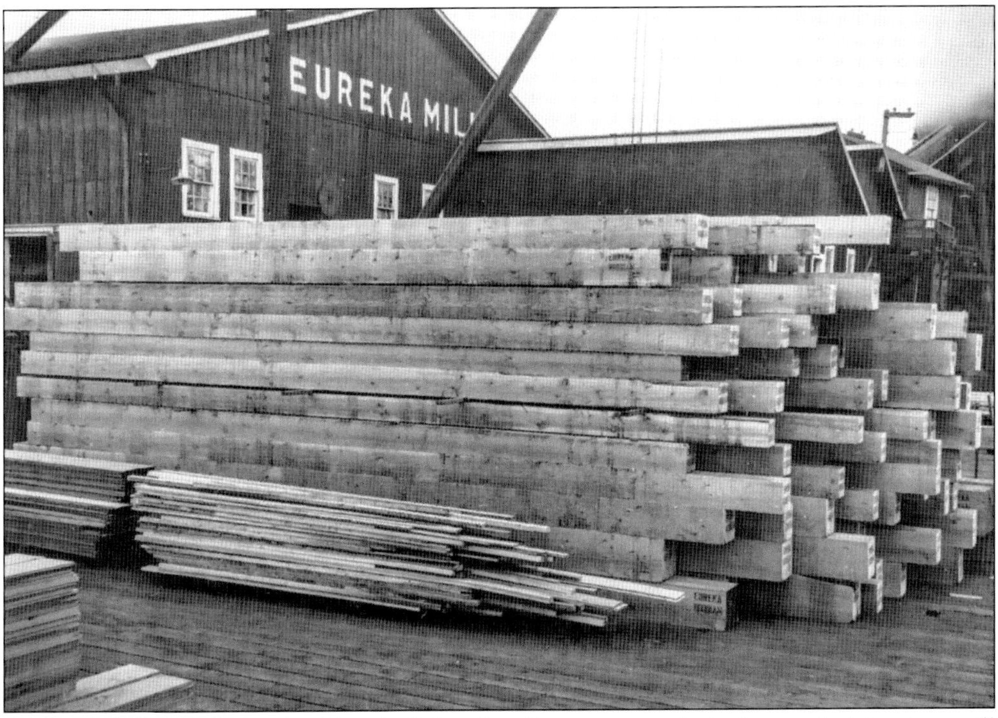

Products of the Eureka Cedar Shingle & Lumber Company are air-drying on the Hoquiam mill wharf. This photograph illustrates the range of dimensional products produced daily at the mill, established by the Polson Brothers in 1910. Its products were shipped to the South Seas, the East Coast, and along the Pacific coasts of North and South America. (AMH.)

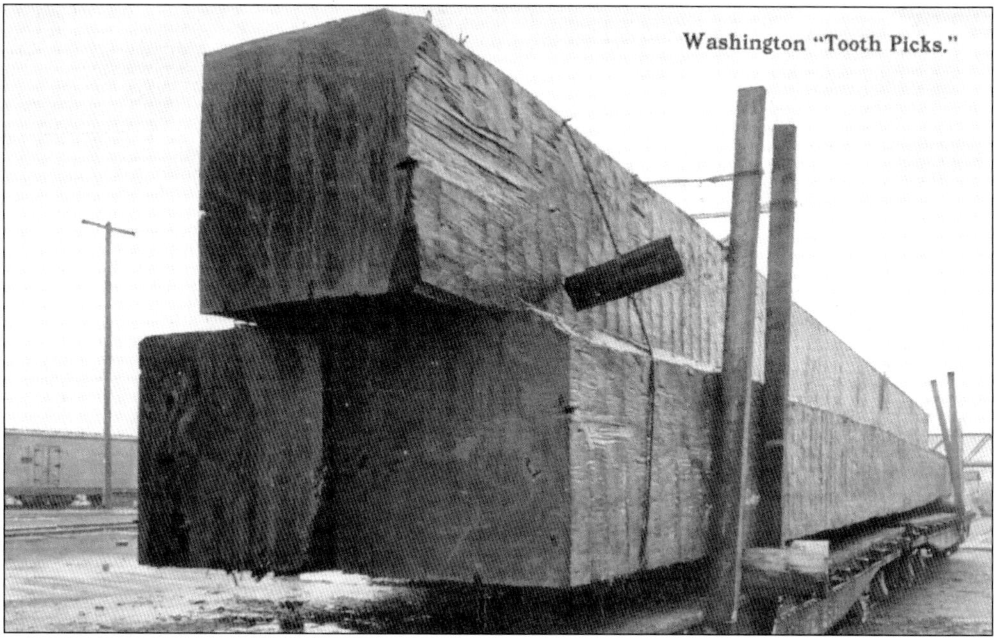

"Washington toothpicks" were more commonly known by the now-insensitive term "Jap squares." These 28-by-28-inch and 30-by-30-inch cedar and spruce cants were produced for the Japanese market from the mid-1910s through the 1930s. They were easy to stow in ships and could be milled to any size upon reaching their destination. (CHSCWC.)

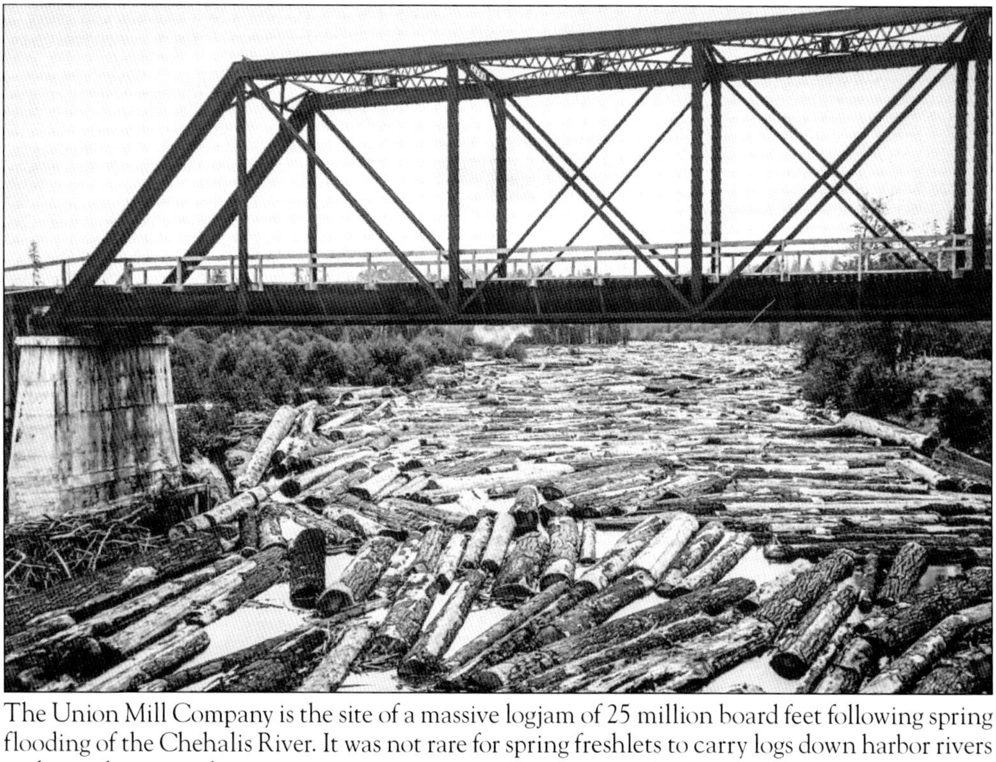

The Union Mill Company is the site of a massive logjam of 25 million board feet following spring flooding of the Chehalis River. It was not rare for spring freshlets to carry logs down harbor rivers and straight out to the ocean, putting a serious crimp into production and profits. (PM.)

The scale of massive, clear-cut logging is displayed in this pre-1920s photograph of the Walville Lumber Company. Subsequent forest regulations have been formulated to prevent cutting near streams and creeks in watersheds and requiring replanting and providing for fish protection. (PM.)

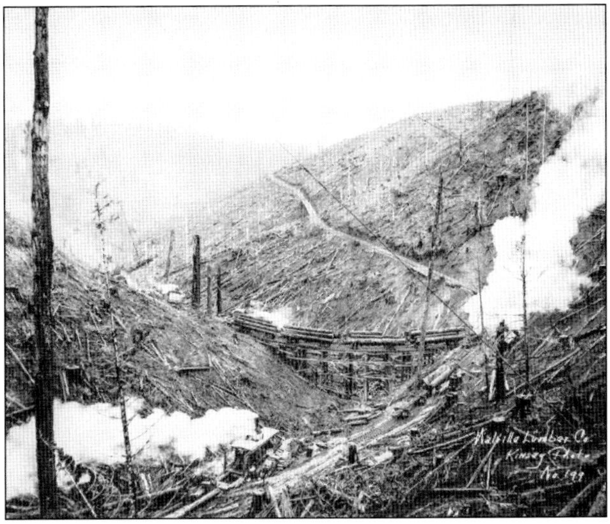

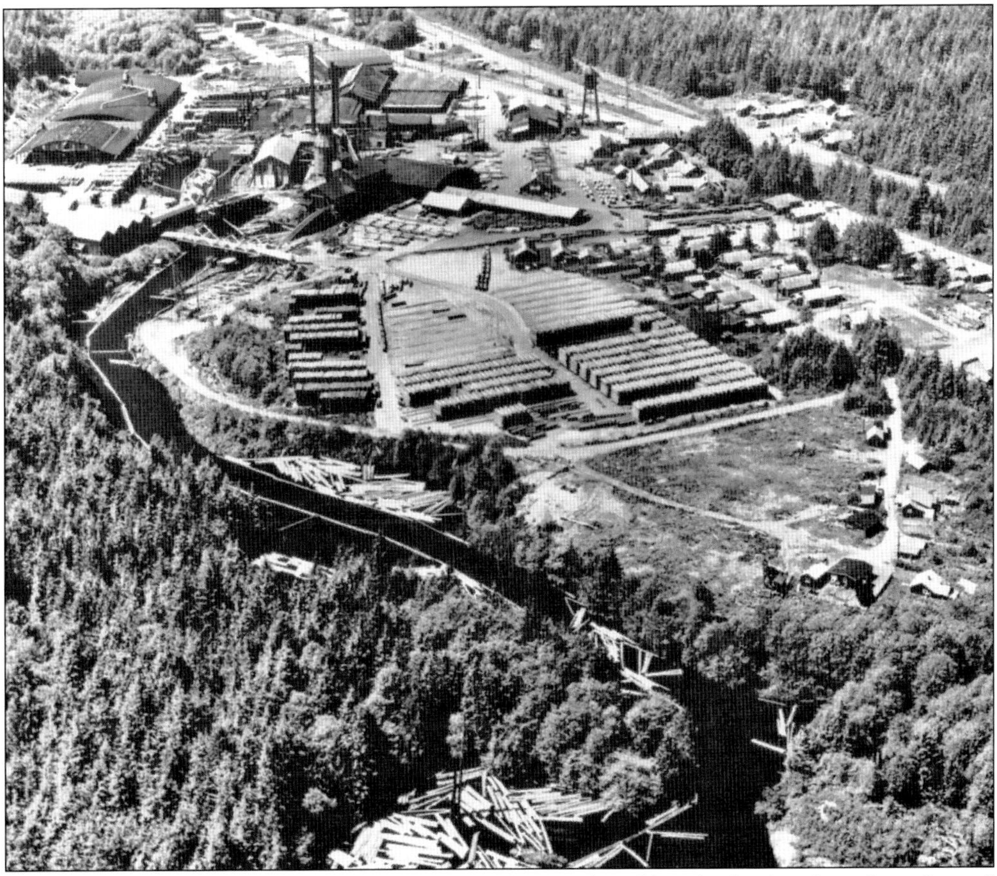

The M.R. Smith Mill & Shake Company at Moclips employed hundreds of workers. Its industrial complex was surrounded by homes, a school, a theater, a community hall, a post office, a mercantile, and a tavern. Under Stu Ferguson, in 1964, the company furnished 27 percent of all US shake production. After 15 years of logging 300 million feet of timber on the Quinault Reservation, the company pulled 13 miles of rail and quit all woods operations. (CHSGWC.)

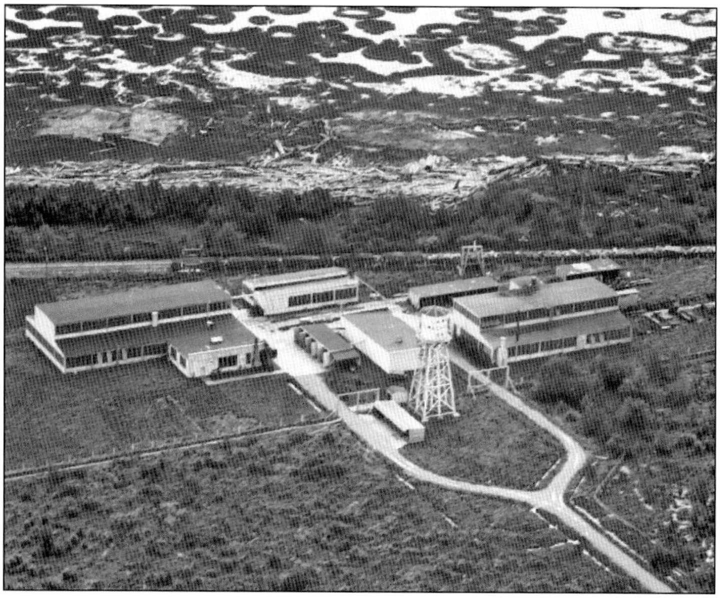

The first Lamb Grays Harbor Machinery Company plant, located close to the Hoquiam River tidewater, is seen here around 1920. It later became a multinational corporation dealing in everything from military production during the world wars, to producing logging equipment, to developing computerized papermaking robots. After 89 years of operation, it closed on June 25, 2001. (PM.)

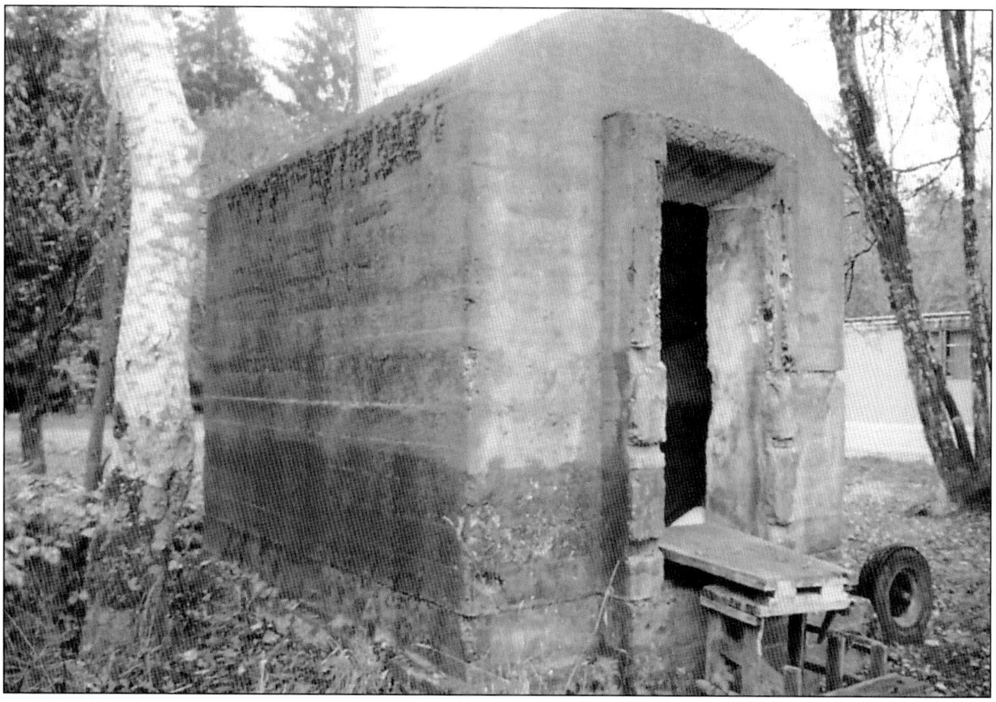

Like an ancient Aztec site, the remains of the Carlisle Timber Company lie hidden in the north woods of Grays Harbor. This massive vault bears witness to the amount of money needed for payroll and for banking in central Grays Harbor. The company was founded in 1912 by W.A. Carlisle after he built his Onalaska Carlisle-Pennell Lumber Company. By 1920, it housed 400 people. A nearby neighbor was the community of Sternsville, named after Josiah Sterns, who operated the Sterns Lumber Co. from 1926 to 1930. (Gene Woodwick.)

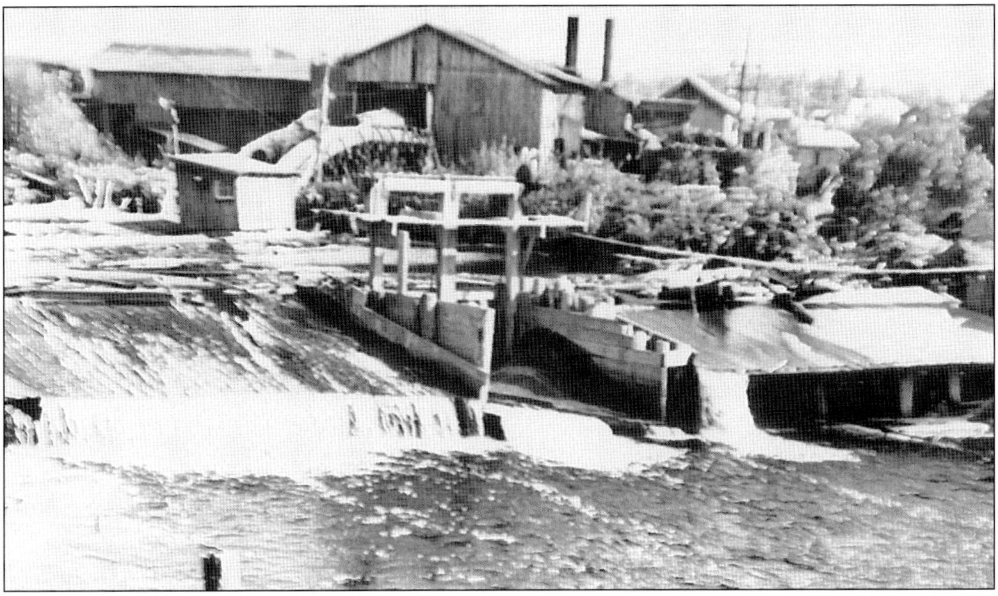

By 1967, the third generation of Pickett, Jackson, Nordman, Baller, Matson, Stangler, and Yales families worked at the M.R. Smith Shingle Mill, producing millions of red cedar products. In 1915, the shingle weavers organized the WA-OR Shingle Weavers Council and plagued the mill with a huge strike. Higher wages were seen in 1916, due to the movement of cedar logs to Japan after the 1923 earthquake. Another major strike occurred in 1930. (CHSGWC.)

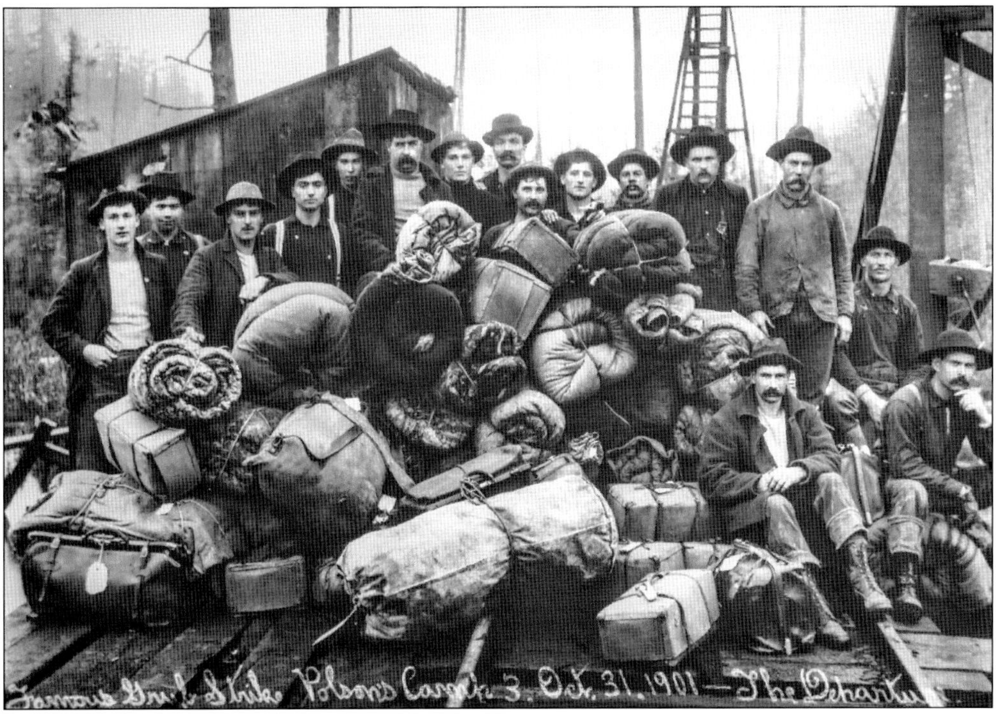

This photograph memorialized the "Great Hunger Strike" of October 31, 1901. It was caused by anger with the "gut buster's" (cook) provender and the horrible living conditions. The entire crew packed up their bedrolls and gear and headed for town. (PM.)

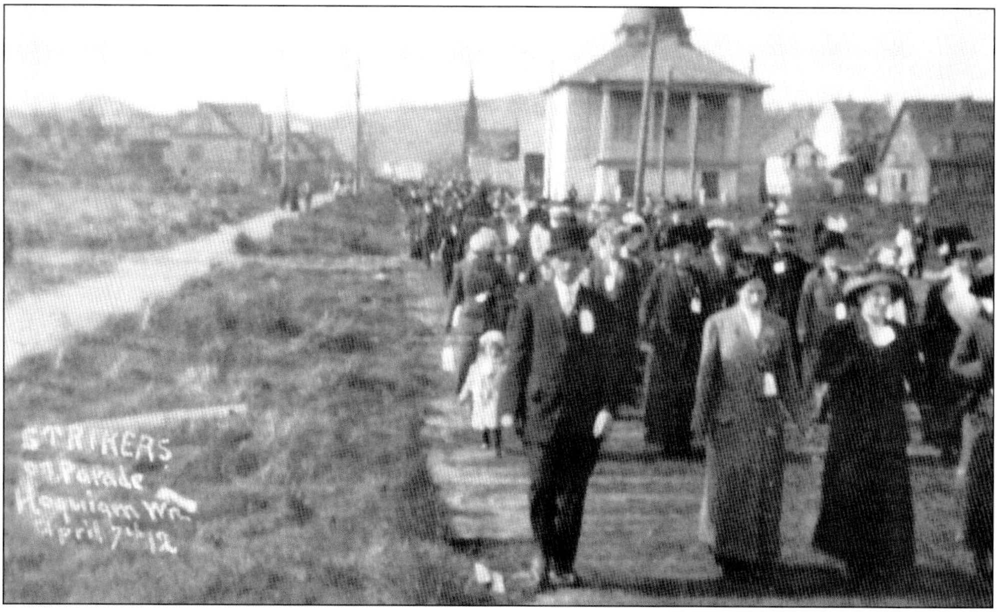

In the early part of the 20th century, 100,000 lumber workers demanded admittance to the AFL Northwest Council of Sawmill and Timberworkers. On April 7, 1912, Finnish socialists from the IWA and AFL marched down Seventh and H Streets in Hoquiam (pictured). In May 1935, a total of 30,000 workers went on strike. Negotiations failed, and on July 8, some 6,000 to 10,000 demonstrators marched from Aberdeen to Hoquiam. In the spring of 1936, approximately 70,000 lumber workers were affiliated with the AWA-CIO International Woodworkers of America, an outgrowth of the old IWW movement. (PM.)

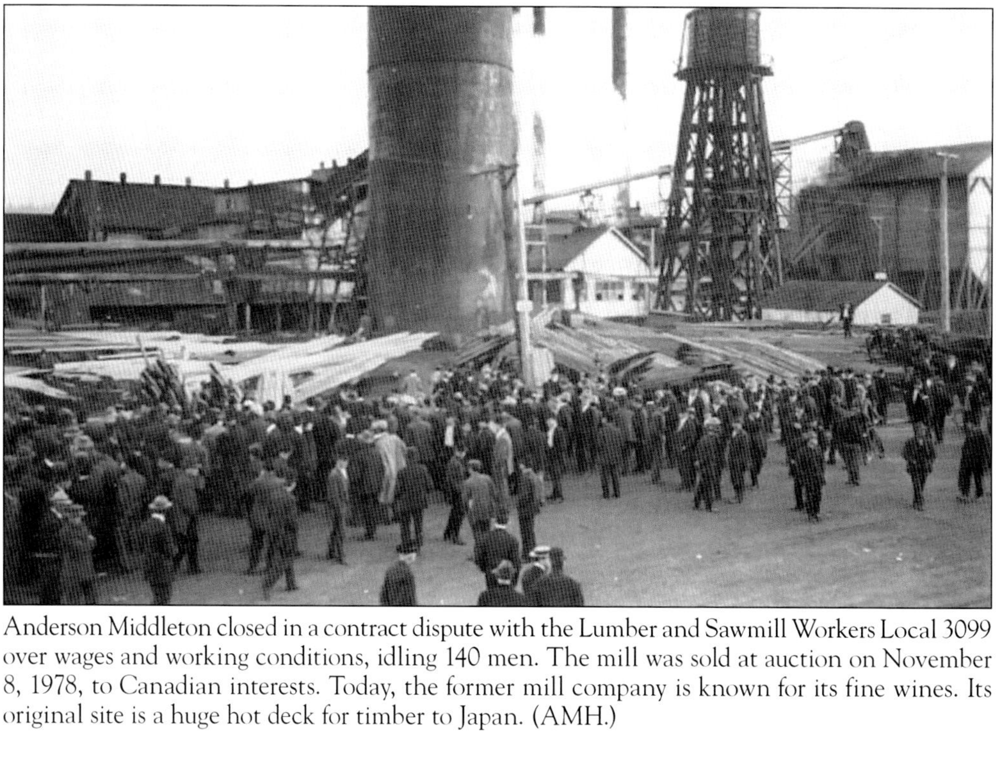

Anderson Middleton closed in a contract dispute with the Lumber and Sawmill Workers Local 3099 over wages and working conditions, idling 140 men. The mill was sold at auction on November 8, 1978, to Canadian interests. Today, the former mill company is known for its fine wines. Its original site is a huge hot deck for timber to Japan. (AMH.)

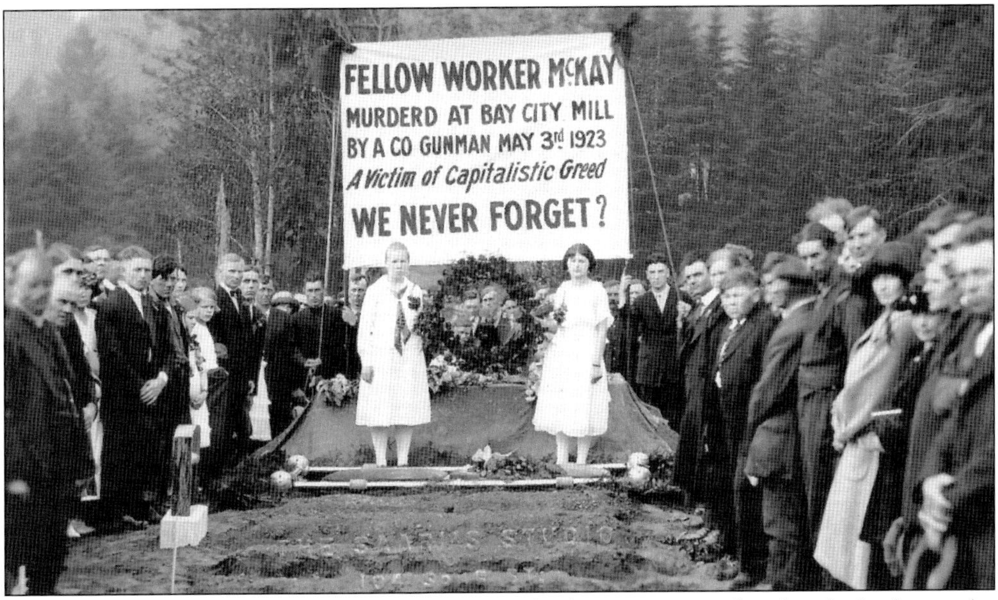

In 1916, IWW membership was a federal crime. Federal troops were ordered to the harbor in the wake of the sawmill strike in July 1917. In the proceedings, Bill McKay was shot in the back of the head. In the 45-year struggle, four Wobblies were executed, one lynched, eight sentenced to prison terms of from 25 to 40 years, four served manslaughter sentences, innumerable numbers were beaten and jailed, and 150 were killed. No employees or agents were ever convicted of murder in clashes with union workers. (CHSGWC.)

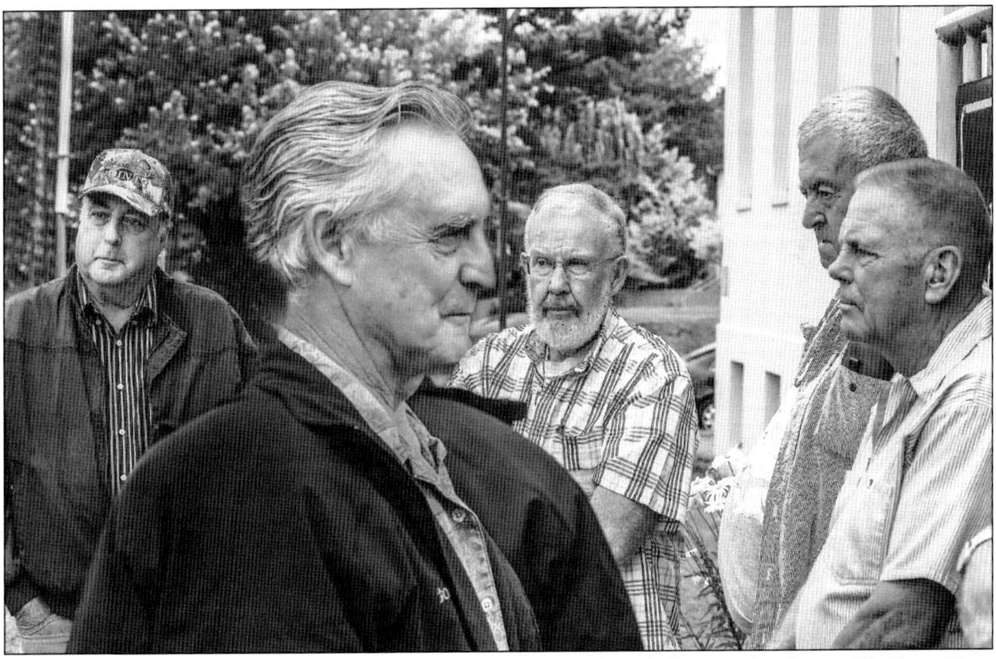

Among those who served the IWA as president were, from left to right, James Phillips, Dan Schlesser, Byron Eager (also the IWA Credit Union manager), Joe Dugay, and Harold Branstad. The rapid decline of companies, with an attendant loss of jobs following the spotted owl decision, led to the demise of the union. (Brian Woodwick.)

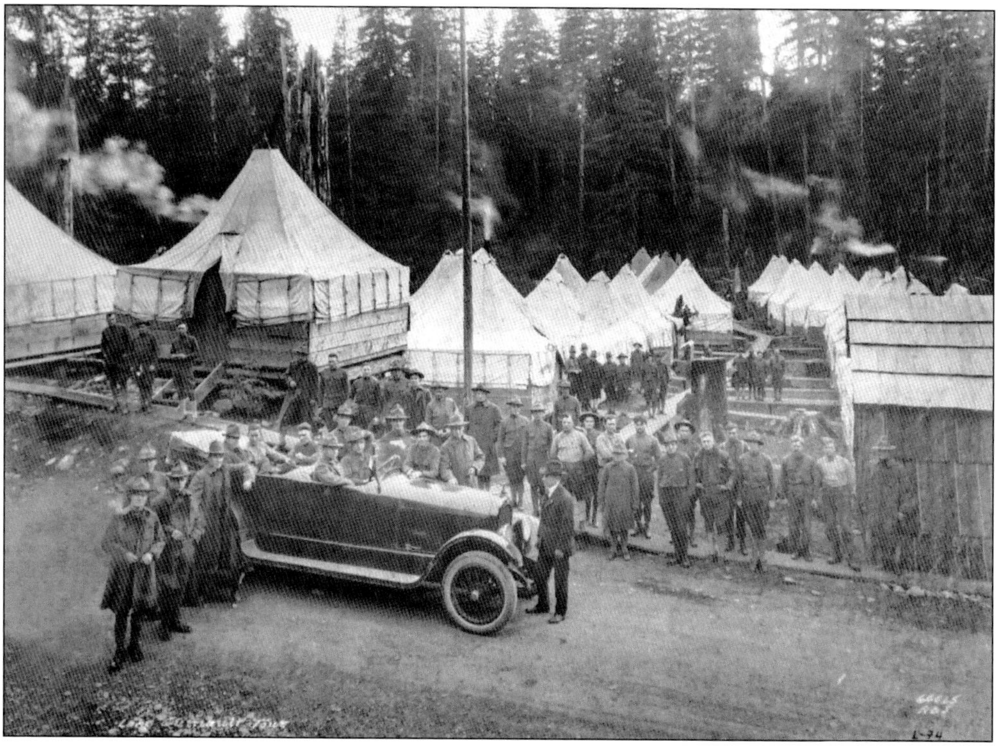

Col. Bryce Disque (in the passenger's seat) and other Army officials inspect the Spruce Boys at their tent camp in the rainy deep woods of the coast. Mobilization of the 10th Engineers, American Expeditionary Force (AEF) consisted of 20,000 troops and 90 sawmills by November 1918. Armistice Day ended the regiment's US Army history. (PM.)

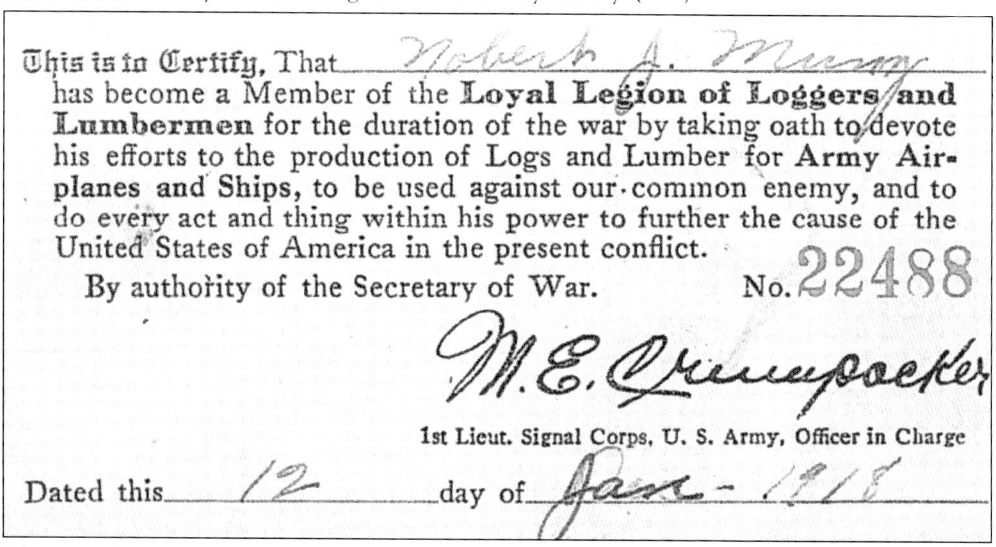

This is to Certify, That Robert J. Murry
has become a Member of the **Loyal Legion of Loggers and Lumbermen** for the duration of the war by taking oath to devote his efforts to the production of Logs and Lumber for **Army Airplanes and Ships**, to be used against our common enemy, and to do every act and thing within his power to further the cause of the United States of America in the present conflict.

By authority of the Secretary of War. No. 22488

M. E. Crumpacker

1st Lieut. Signal Corps, U. S. Army, Officer in Charge

Dated this 12 day of Jan - 1918

This Loyal Legion of Loggers and Lumbermen (Four-L) membership card from World War I, signed by both citizen and soldier, displays the oath. The military sent city boys from New York and New Jersey, who had never seen such woods, to log the rain-soaked coastal forests. It was no wonder that 25 experienced, civilian, local loggers could turn out the same amount of wood as 100 Spruce Division soldiers. (WSM.)

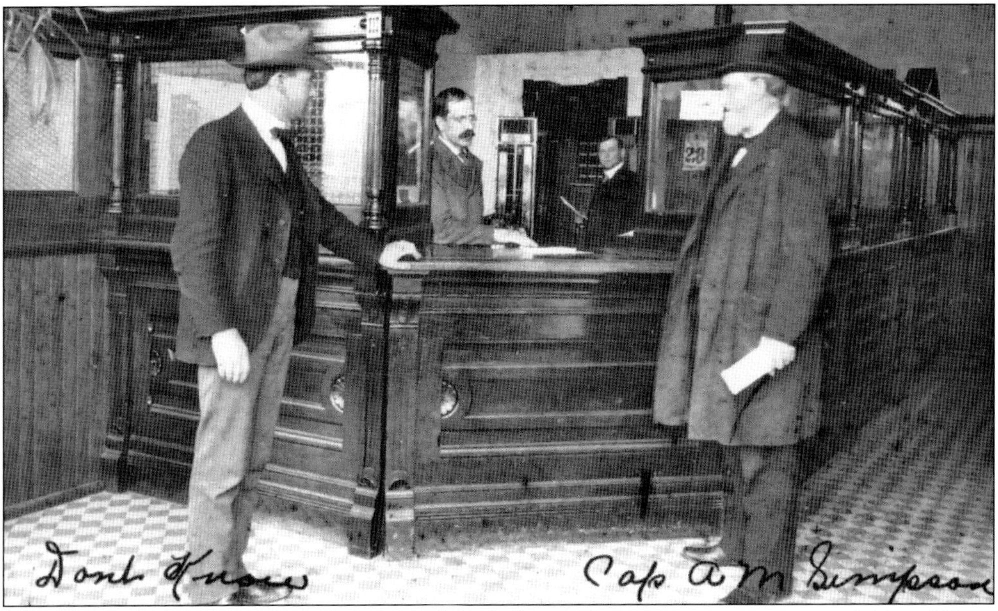

Capt. Asa M. Simpson (right) and an unidentified man discuss money matters with Tony Wirty (behind counter on left) and an unidentified clerk at the Hoquiam National Bank. Simpson incorporated the North Western Mill Company on February 28, 1884, along the banks of the Hoquiam River. After the government, Simpson was the largest landholder on the Olympic Peninsula, with 30,000 acres, including the township of Hoquiam and four miles of shoreline, along with all vessels on the harbor. (WSM.)

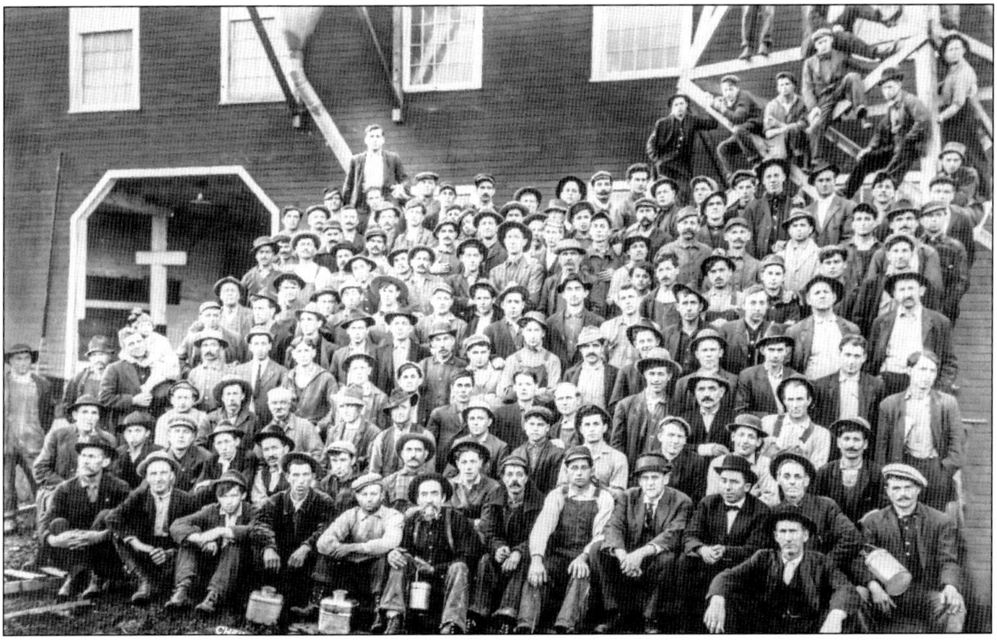

The Chehalis Fir Door Company at McCleary was built in 1911 as "the biggest door factory in the entire world." It employed 300 men who produced 4,500 doors a day. The company went bankrupt in 1907 and was acquired by Henry McCleary. The firm set a one-day record on December 31, 1950, with 1,002,499 doors produced. (PM.)

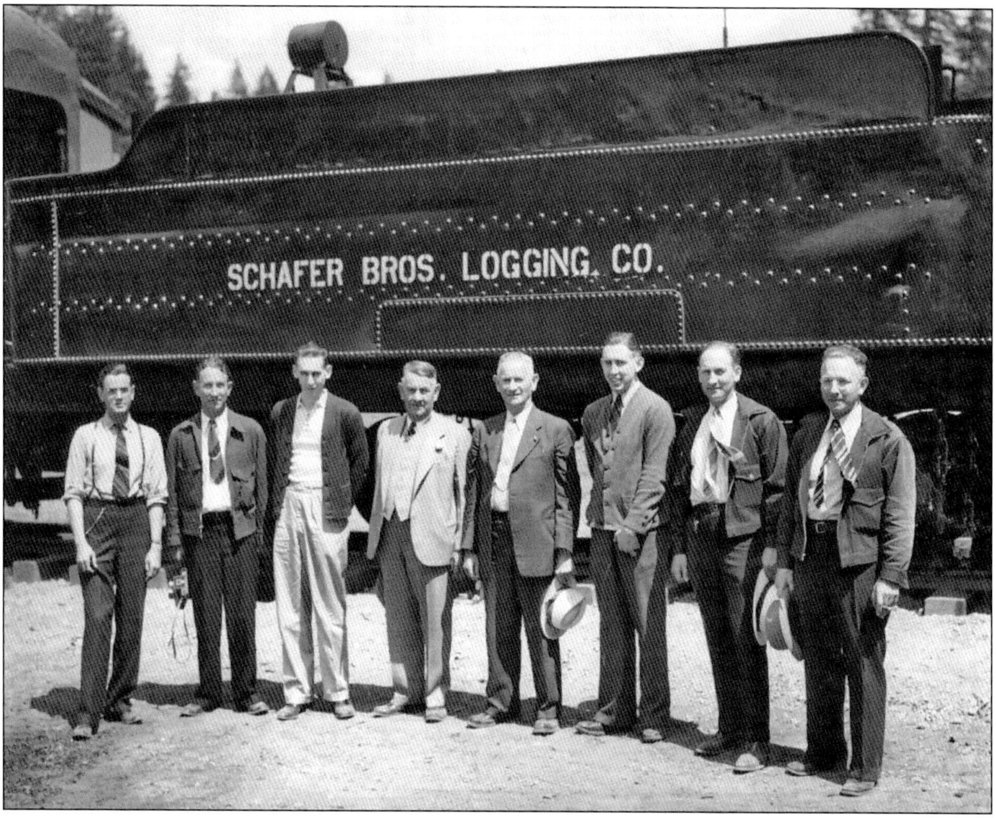

Schafer Brothers Logging Company began in 1893, when brothers Peter, Albert, and Hubert Schafer began logging on the family's Satsop homestead. In 1913, they bought a 45-ton Heisler locomotive to begin their railroad logging operation. At its height, Schafer Brothers, with 3,000 employees, was one of the largest logging, milling, and shipping companies in the Pacific Northwest lumber industry. The second generation of Schafers sold the firm to the Simpson Timber Company in 1955. (AMH.)

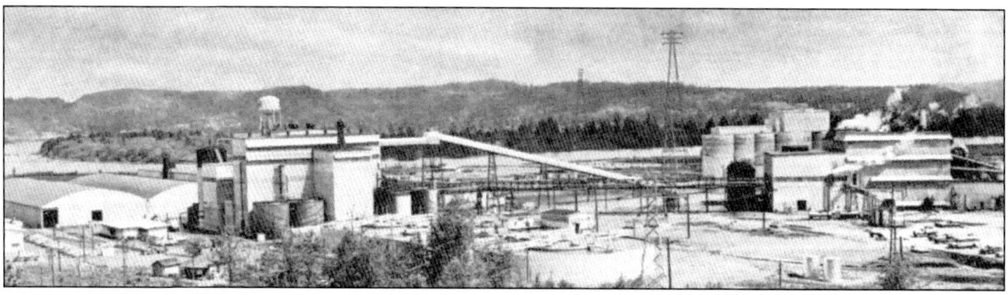

The Cosmopolis Weyerhaeuser sulphide mill complex was built on the 60-acre site of the old Grays Harbor Commercial Company. It began production in 1957 with capacity of 400 tons daily. The 17 units of the complex were later sold to the Cosmos Fiber Company. (CHSGWC.)

Seven

High Lead and High Climbers

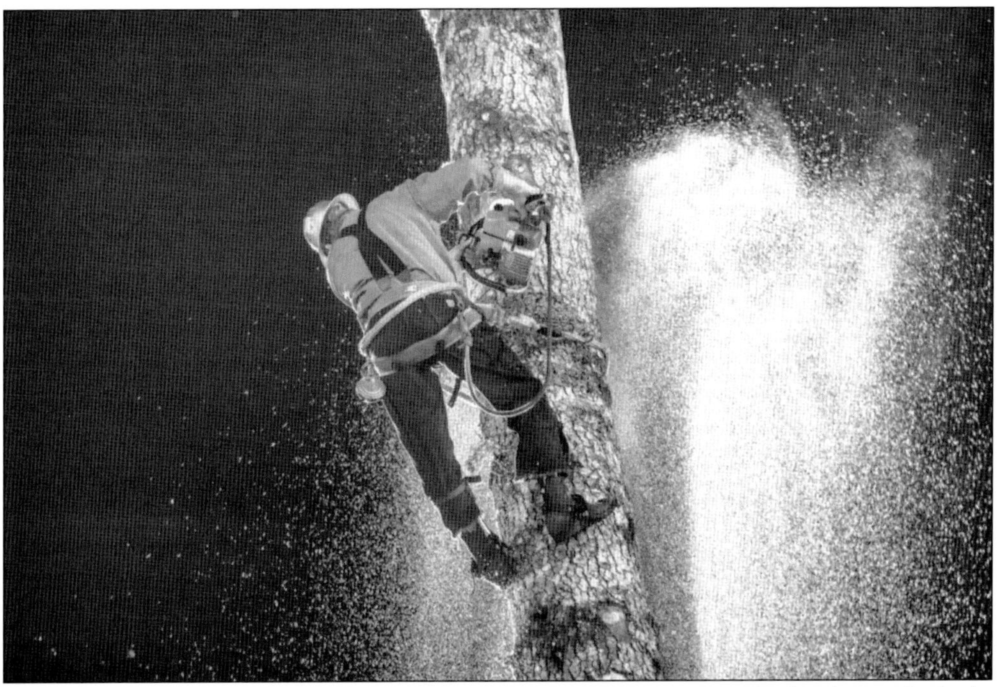

John Wayne Rodius thrills the crowd at the annual Hoquiam Logger's Playday during the high-climbing/tree-topping event. This annual exhibition of strength, skill, and courage always draws the attention of the crowd, from the time they see the climber strap on climbing spurs and the topping power saw swinging from his belt, to the slacking of his rope as he gets ready to top the "tree." (Brian Woodwick.)

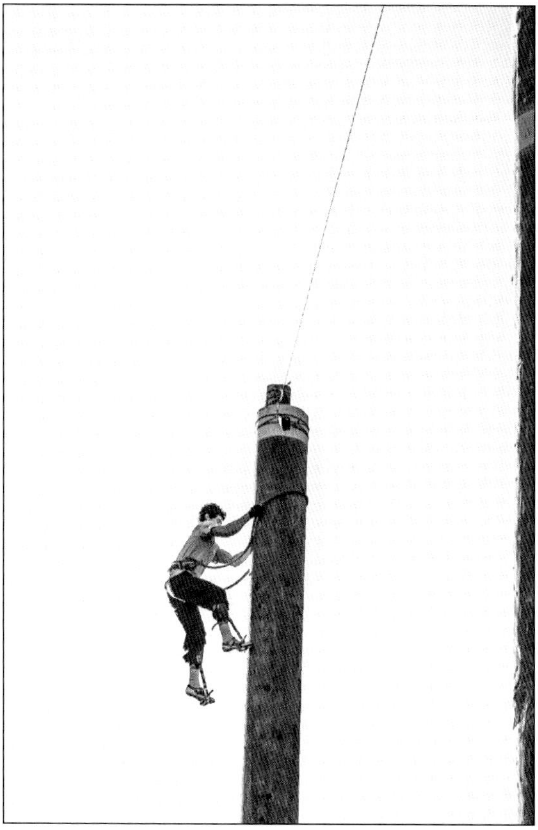

Brian Bartow nears the top while competing in speed climbing. Participants aimed for the fastest time from the ground to the top and back. Flipping the 7/8-inch wire rope with its steel core is not only a test of muscle strength. The spurred boots add weight to the legs in this strenuous event. The descent is amazing, when the climbers pull away from the tree, touching a spur every 15 feet as they plummet to the ground. (Brian Woodwick.)

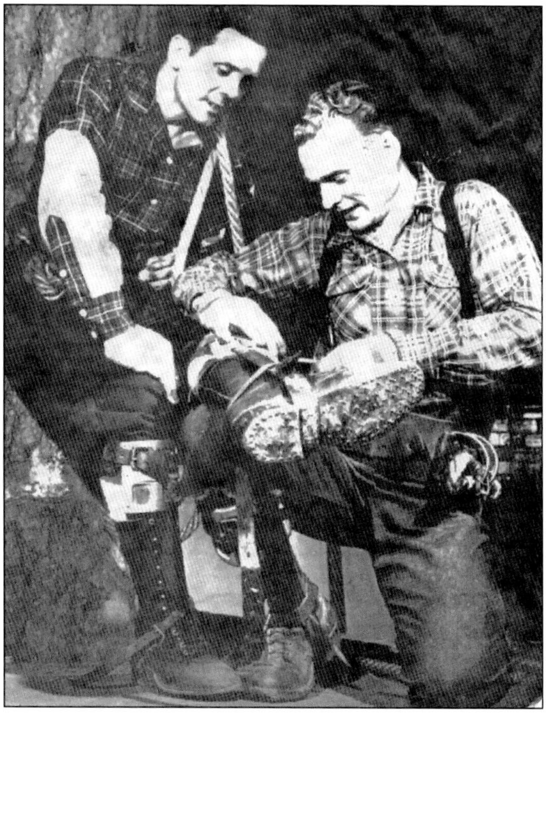

Multiple all-around world champion and record holder Harold "Hap" Johnson buckles on Mel Harper's tree-climbing spurs. Both men were well known for their championship performances, both on and off the job. Johnson won the World's Topping Championship a record 14 times (1947–1973) and the all-around trophy three times—pretty good for a man with three kids who worked in the woods every day. He retired in 1984 after performing in logging contests for 35 years. (CHSGWC.)

Clyde "Bing" Smith, a high rigger (climber) for Rayonier, was held by a steel-covered rope to trim branches from the tree trunk. He then topped the tree at 100 feet or above in order to rig blocks and receive lines forming the basis of a spar tree at yarding and logging sites. The fully rigged spar became the basic high lead operation tool for logging area landing sites from 1910 to 1912. (CHSGWC.)

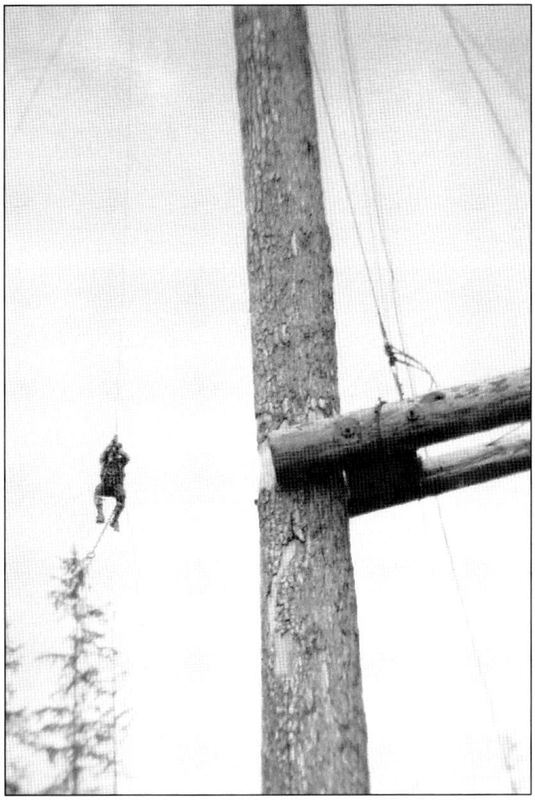

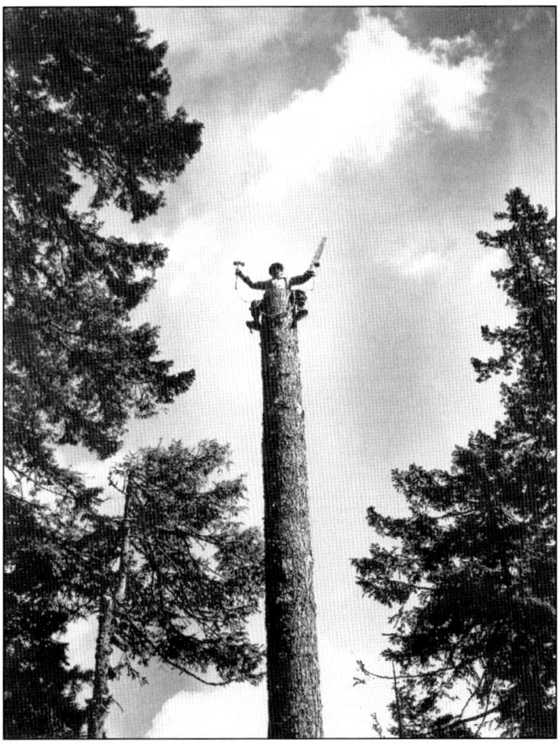

Bobby Sholes of Aberdeen waves to indicate success in removing the crown of a spar tree at a Sholes-Lutton Company site. At the topping point, the tree towers 40 to 60 feet above the high climber, wielding a 45-inch bar plus the engine on the power saw. A climber would stand upright on the fresh cut to ride the spar as it swung in a 25-to-30-foot arc. Sholes is the model for the Lawrence McFeron sculpture at Aberdeen Morrison Park. (CHSGWC.)

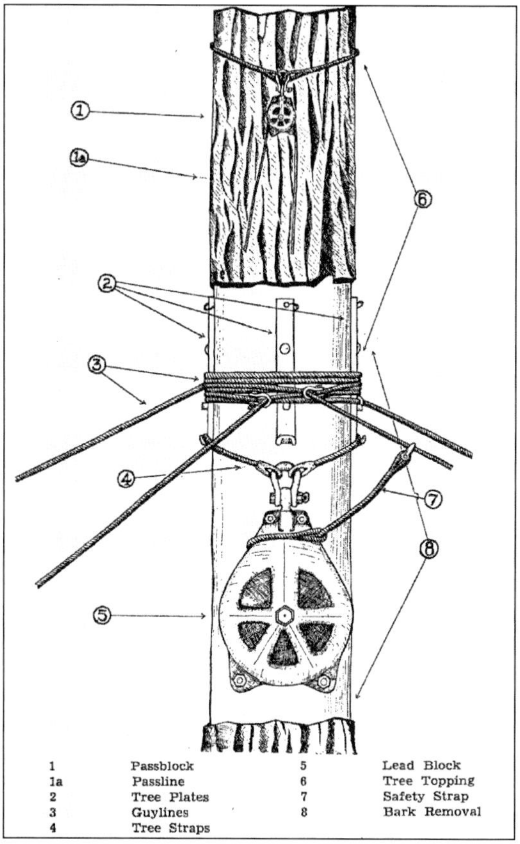

Rigging a bull block for a spar was quite a feat, as shown in this *Logger's Safety Standard* illustration. The safety manual was invaluable for keeping abreast of the latest in logging equipment operations. Today, the manuals display the work of days gone by. (CHSGWC.)

The Skagit Steel Spar replaced the tree spar in the late 1960s. It is seen here at an operation on the Quinault Indian Nation reservation lands. The movable spar saved time and money for the logging companies, although it eliminated some crew members. (QDNR.)

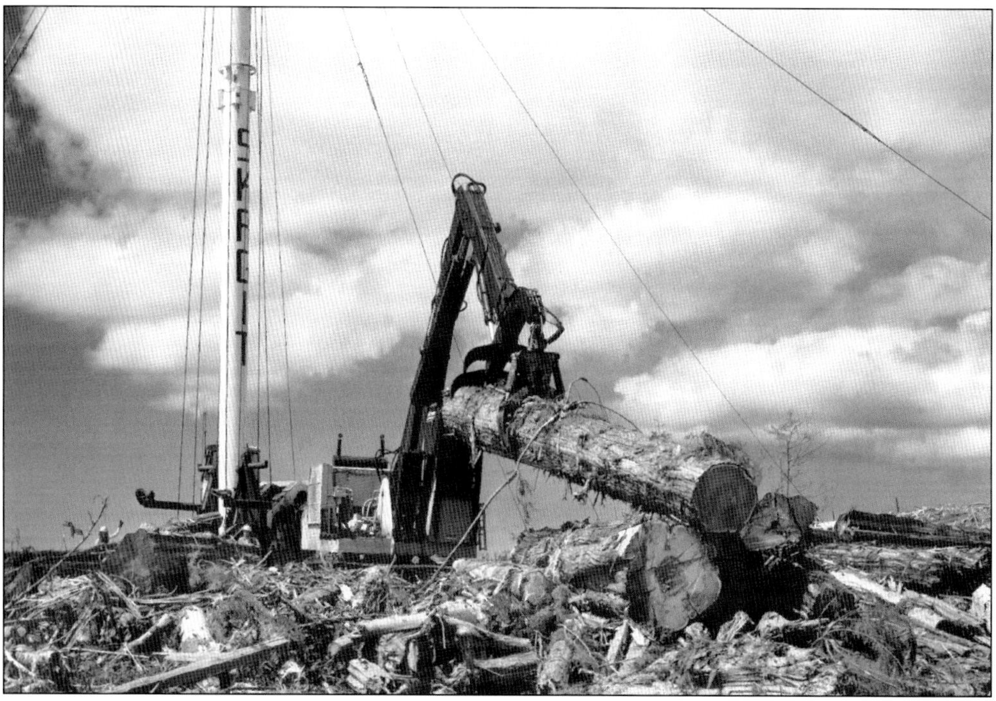

In this operation, a "hayrack" boom lifts an entire load from a truck to a railcar. Today, a similar device lifts loads from trucks for transport within the burgeoning industry of chip production. The technology reduces thousands of logs a day for transporting to paper companies and other timber-product entities. (PM.)

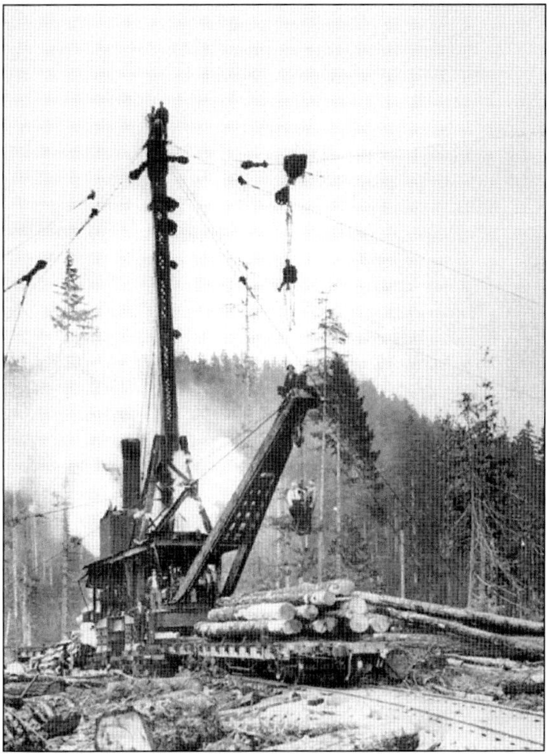

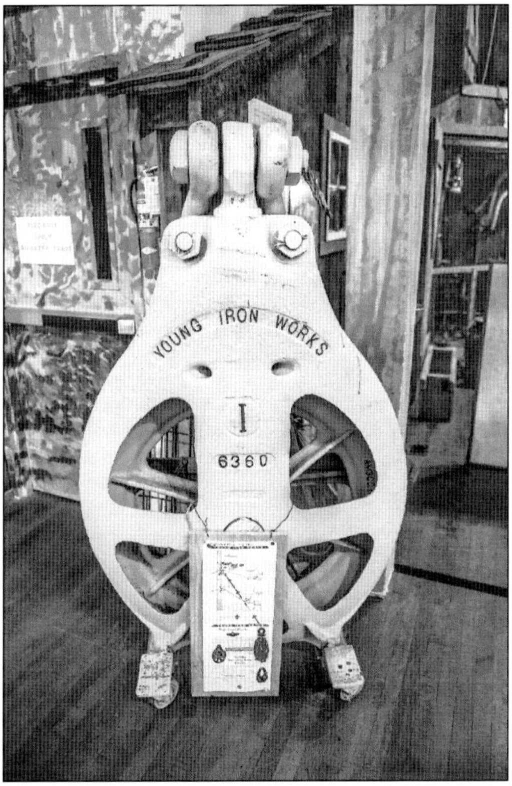

This shoulder-high Young bull block, on display at the Aberdeen Museum of History, indicates the size of timber harvested for nearly a century in Grays Harbor. Rigging slingers put the blocks into place at logging operations. (Brian Woodwick.)

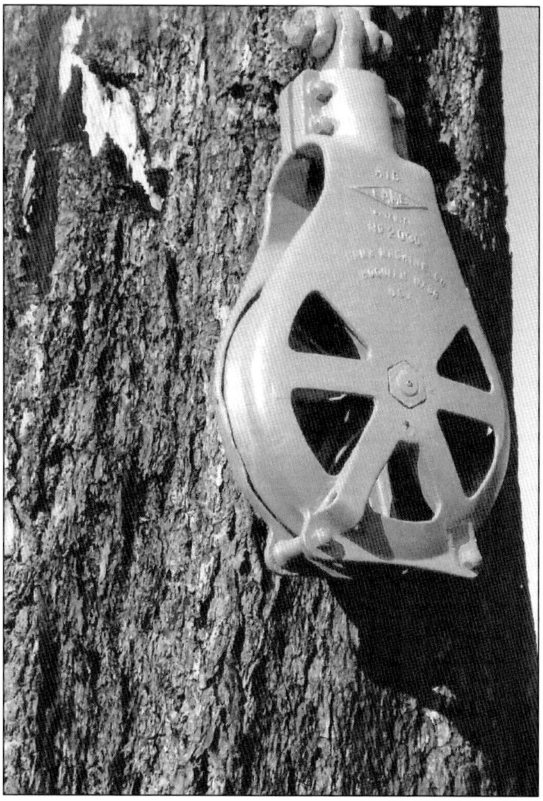

In logging country, bull blocks were as common as coffee cups in the harbor's operations. The company grew from a producer and supplier of logging equipment under Frank Lamb into a multinational corporation under his son, Frank Jr. The company ceased operations in 2001. (Gene Woodwick.)

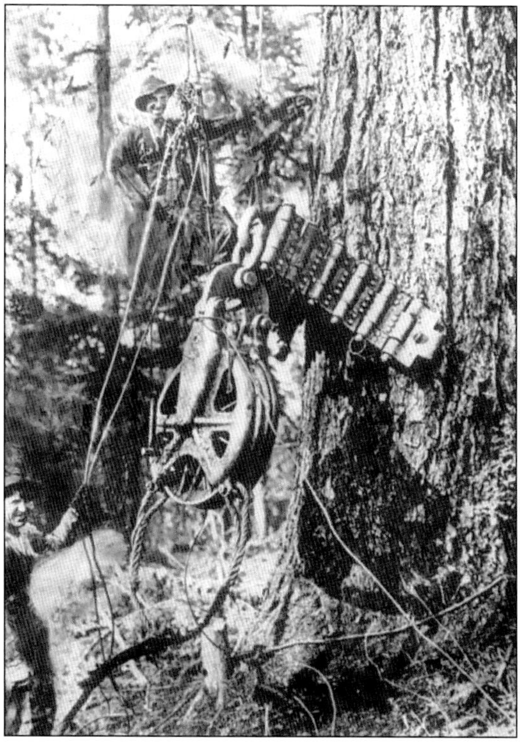

This rare photograph shows just one of numerous logger inventions. The spar tree bull block girdle was short-lived, but it was definitely a unique way to secure the block onto a spar tree. (PM.)

This type of skyline carriage for high-lead logging was affectionately known as "The Bullet." It had a gas pony engine and a brake system. The carriage (or traveling block), used to load logs, rides on skyline cable between the head tree and a stump. This carriage can be seen at the Brooklyn Tavern at the Saginaw Logging Show site at North River. (Gene Woodwick.)

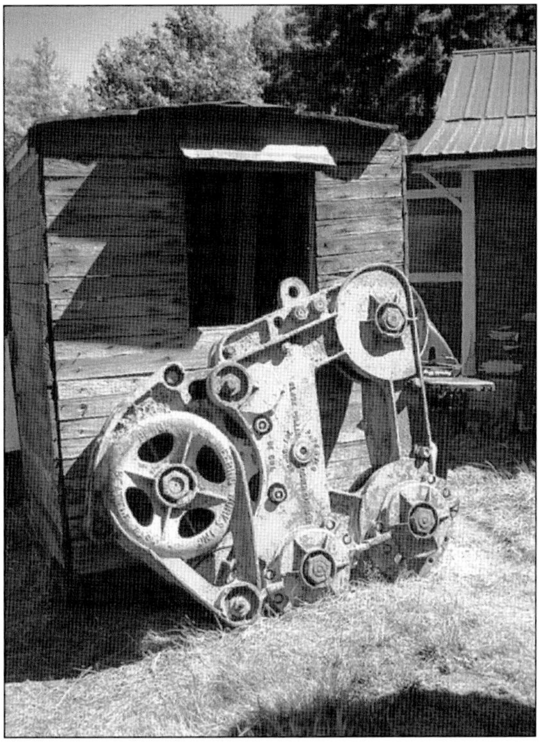

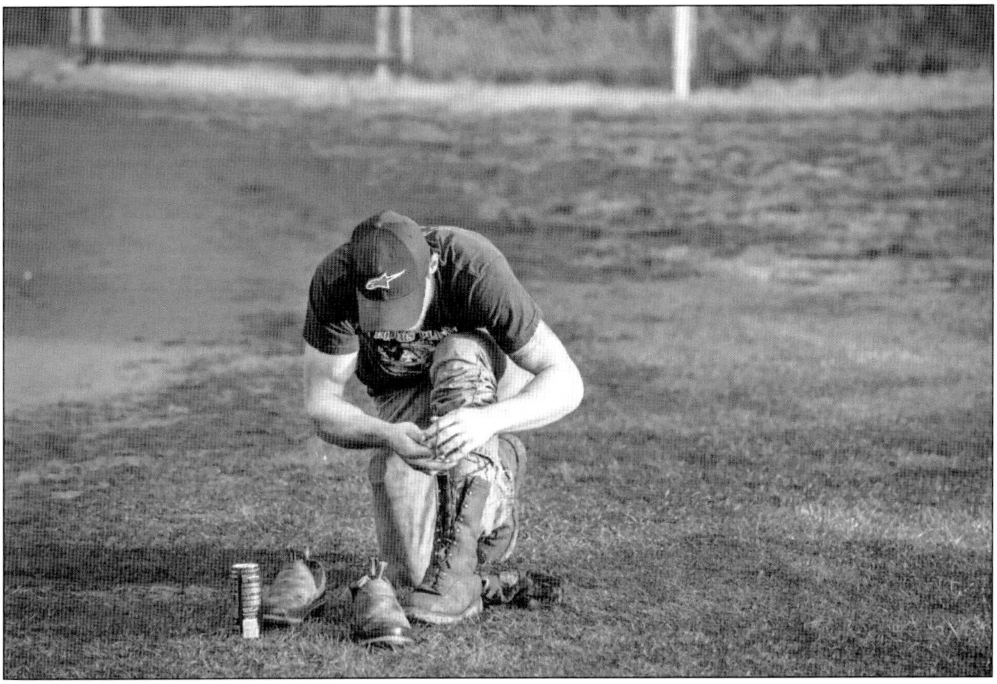

As important as mechanized tools are, loggers pay close attention to their footwear. Customary forms of footwear for coastal loggers are the slip-on shoes known as "Romeos" and the steel-caulked boots called "corks," which reduce injuries and save lives in scrambles over thick bark. Coastal corks sport longer spikes on the soles than do dry-country corks. (Brian Woodwick.)

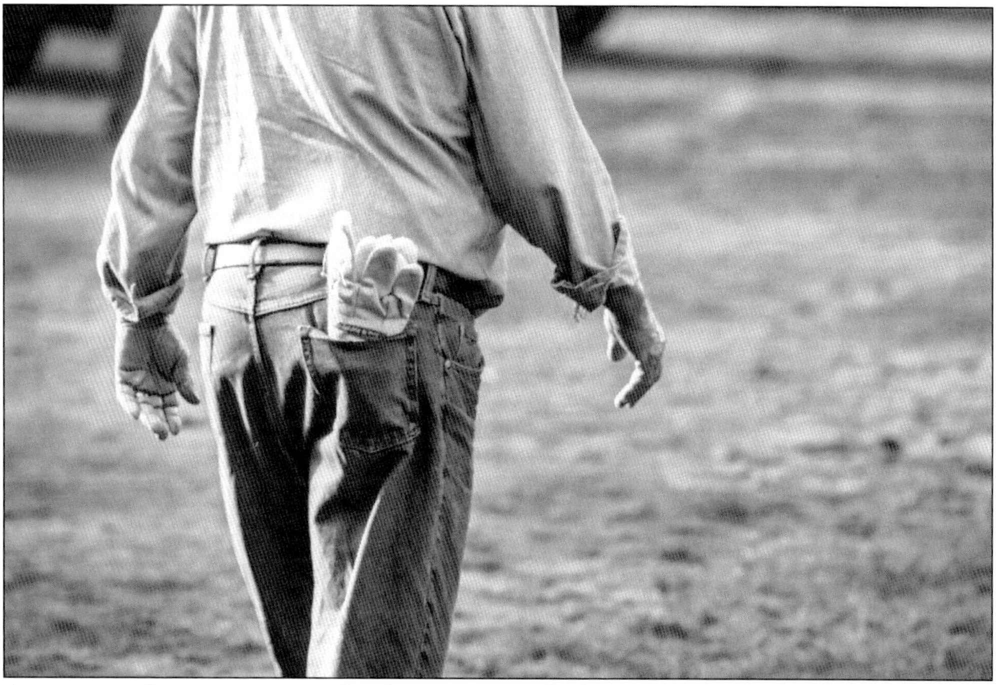

A constant in logging country is the ubiquitous pair of White Ox canvas gloves peeking out of the rear pocket of staged-off jeans. The gloves are bought by the dozens and are used in everything from protecting hands, to wiping off greasy equipment, to smacking dirt off the seat of a truck or Cat. (Brian Woodwick.)

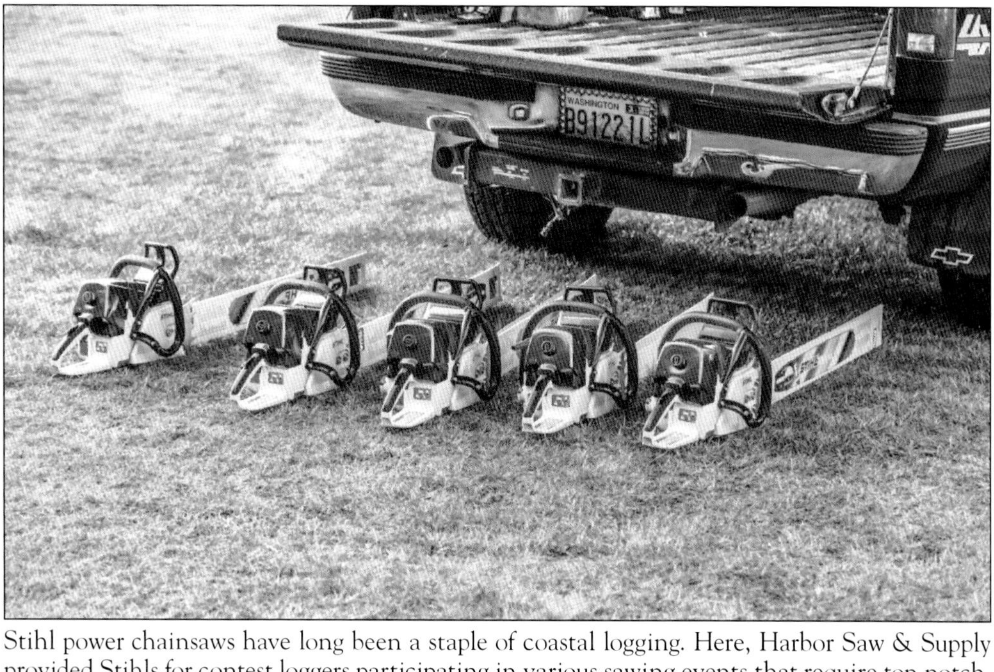

Stihl power chainsaws have long been a staple of coastal logging. Here, Harbor Saw & Supply provided Stihls for contest loggers participating in various sawing events that require top-notch, high-quality saws. Little boys can spot a bushler or a faller by the well-developed muscles on the fingers. (Brian Woodwick.)

Eight

Truckers and Log Movers

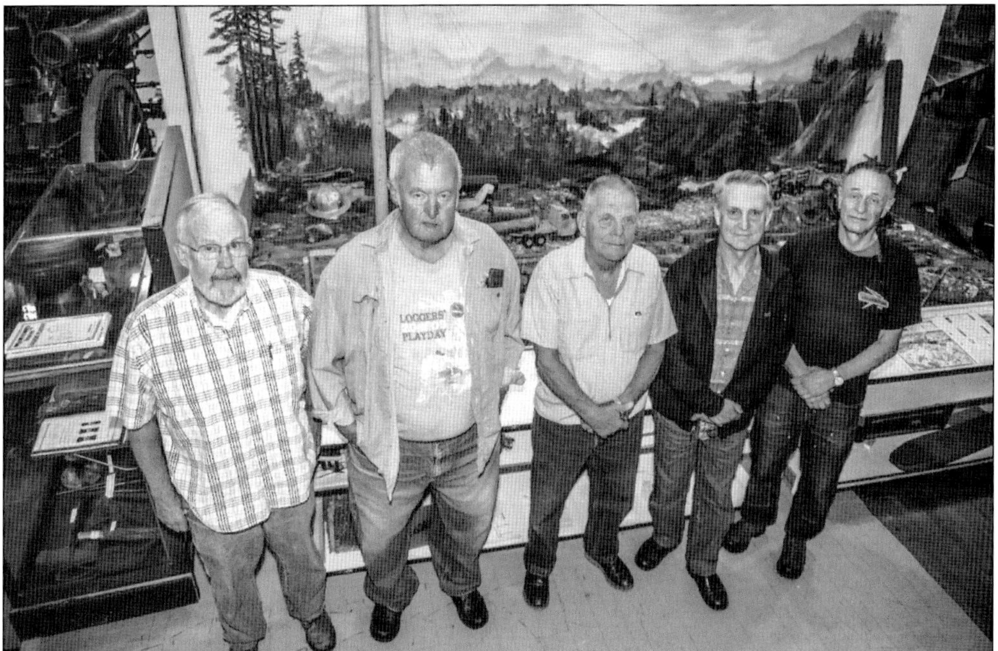

To little boys in Grays Harbor, seeing guys like these in log trucks was like seeing rock stars. Shown here are, from left to right, Byron Eager; Joe Dugay, mechanic, driver, and equipment operator; Harold Branstad, contract logger for Branstad Logging Company; Jim Phillips, Weyco lowboy driver; and Bobby Shoals, Sholes & Lutton high climber. (Brian Woodwick.)

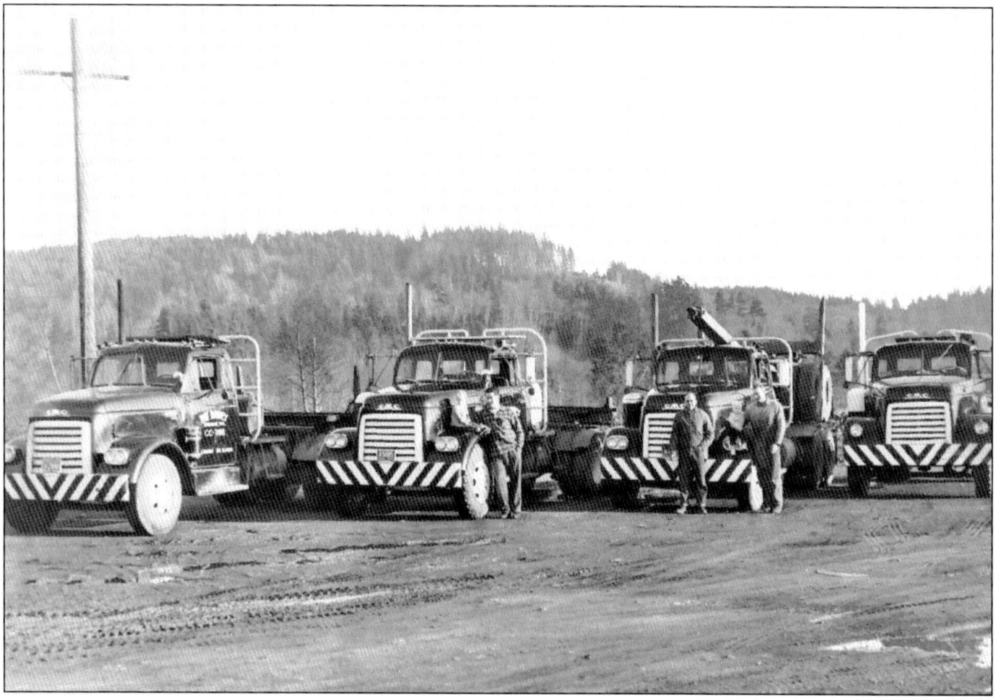

Pat Rabey Trucking, Inc.'s new GMC "Jimmys" provided a photographic opportunity for, from left to right, Patrick, Troy, and David Rabey. (Rabey Trucking Company.)

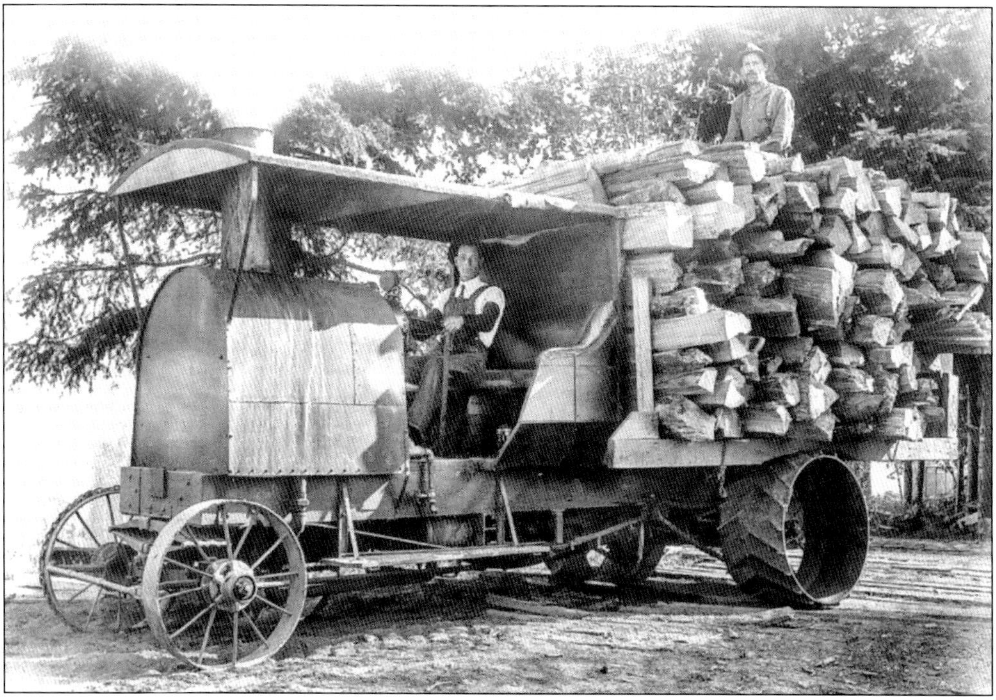

The first logging truck in Grays Harbor was this Nash-Quad truck. It is loaded with veneer blocks on the beach, four and a half miles north of Copalis Crossing. The four-wheel drive was especially good during hauling at high tide. (CHSGWC.)

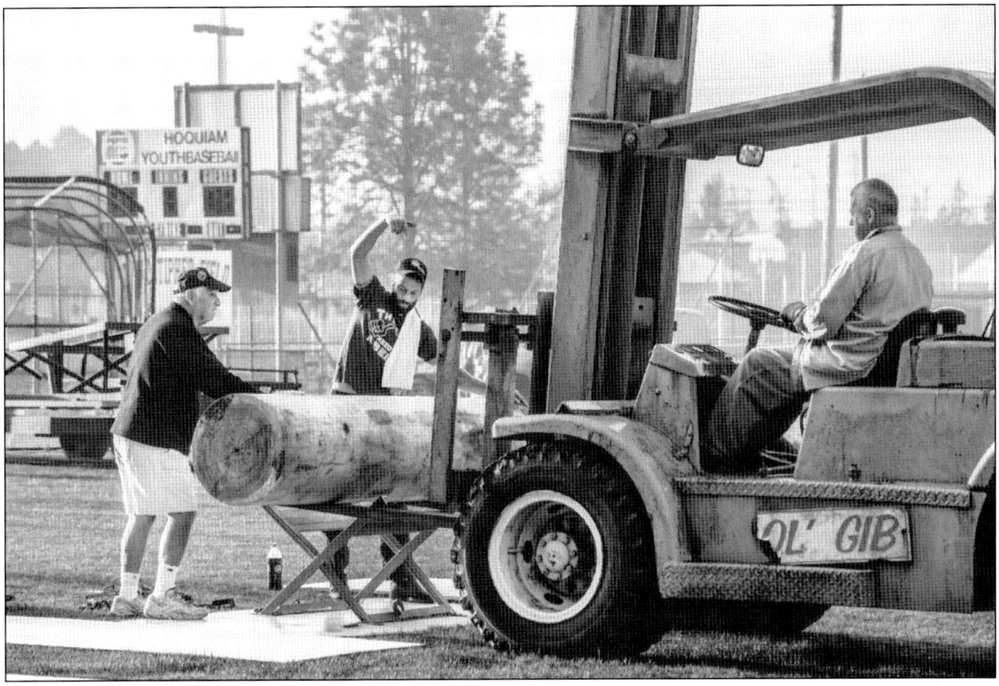

Trucks and equipment are a big part of loggers' lives. Here, "Old Gib," a Heisler, is being driven by Joe Dugay who held various logging jobs from 1957 to 2000 in Grays Harbor. (Brian Woodwick.)

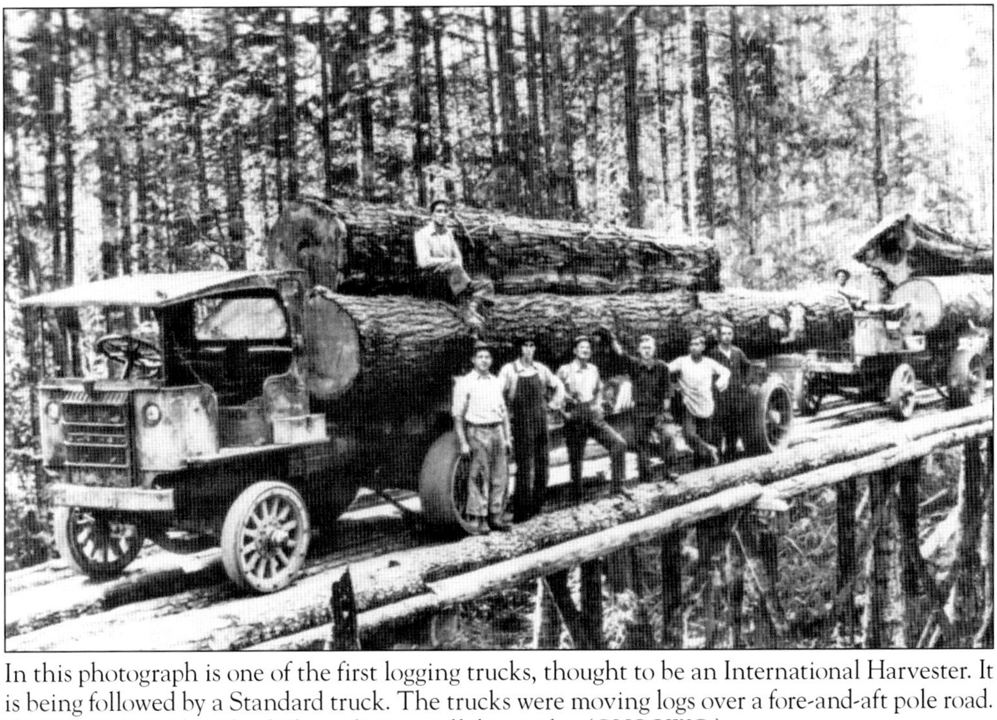

In this photograph is one of the first logging trucks, thought to be an International Harvester. It is being followed by a Standard truck. The trucks were moving logs over a fore-and-aft pole road. Sheer rails kept the wheels from slipping off the tracks. (CHSGWC.)

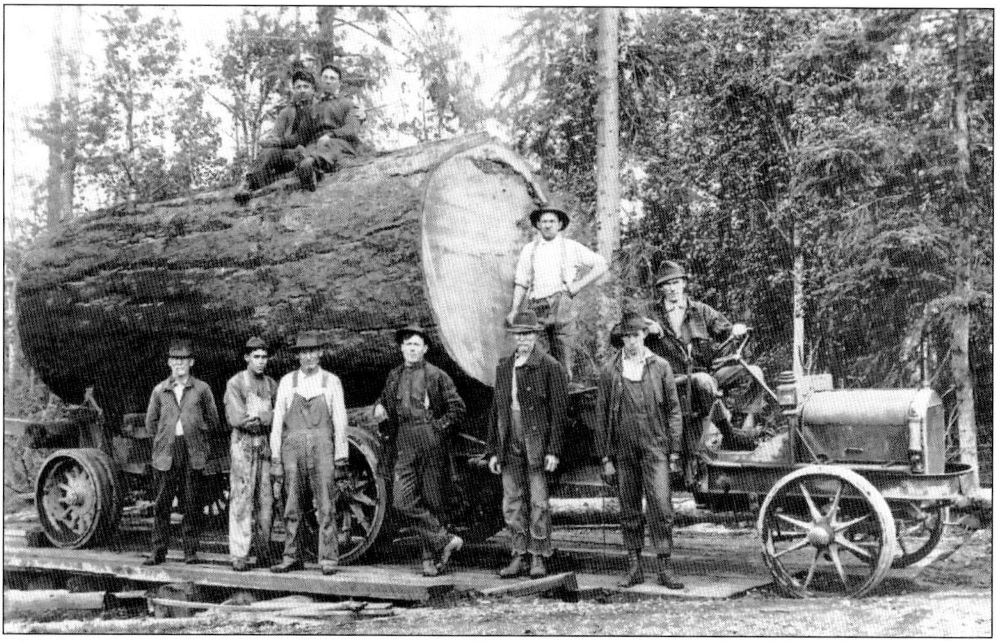

This White truck, with its pre-pneumatic tires, was considered to be a sturdy hauler. Note the acetylene light on the cab. (CHSGWC.)

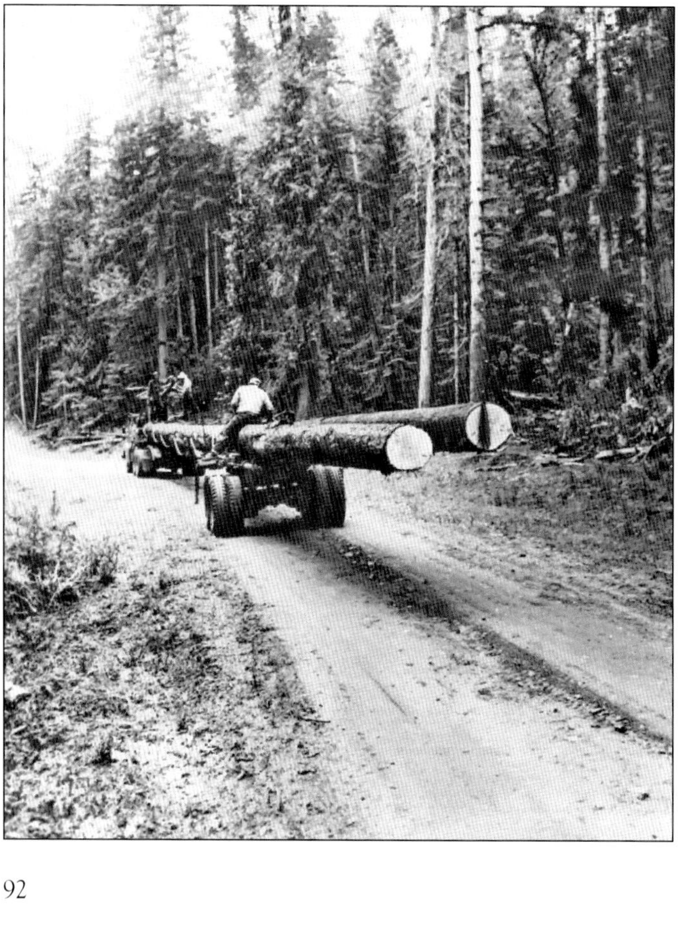

A logger acts as spotter for the driver of an early tractor, pulling loads mounted on truks as a precursor to today's low boys. (CHSGWC.)

Don Bell checks this big stick, loaded onto his truck instead of on a low-boy. (PM.)

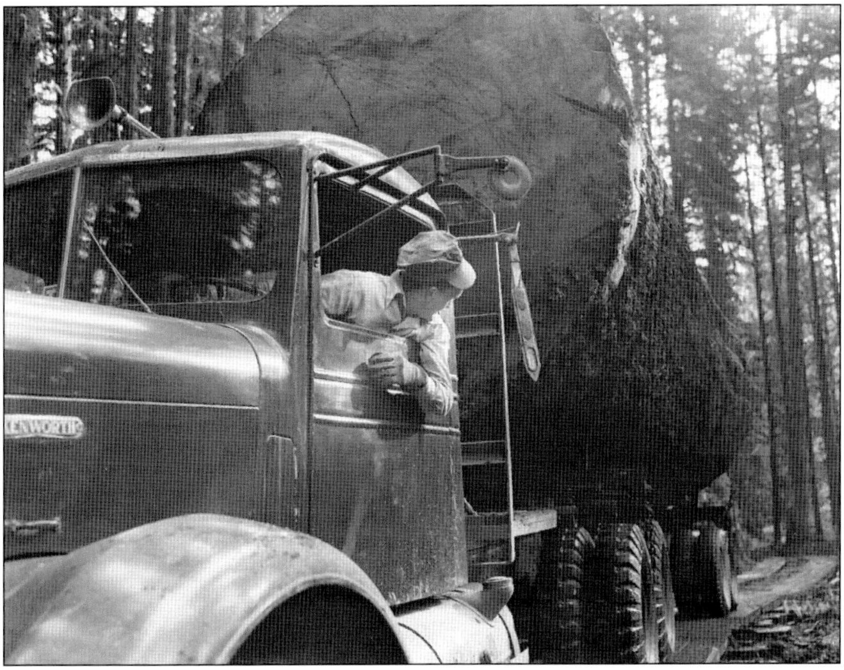

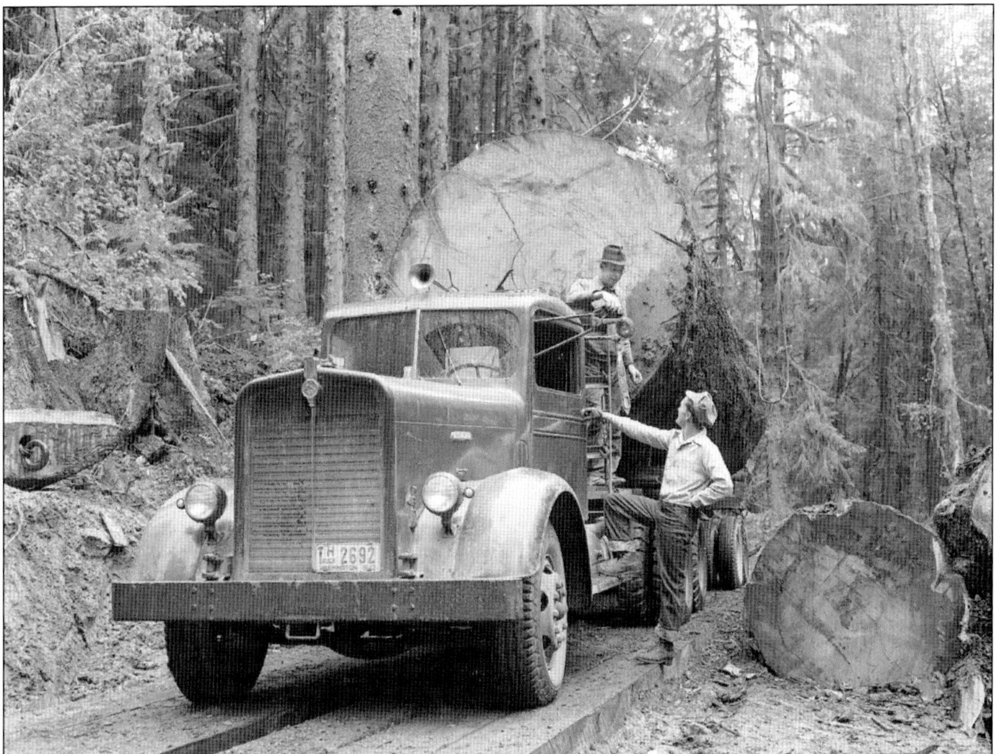

Don Bell never believed that this one-load log would get the truck stuck in the roadbed. In addition, the log ended up whacking the North River bridge superstructure as a state trooper came into view. (PM.)

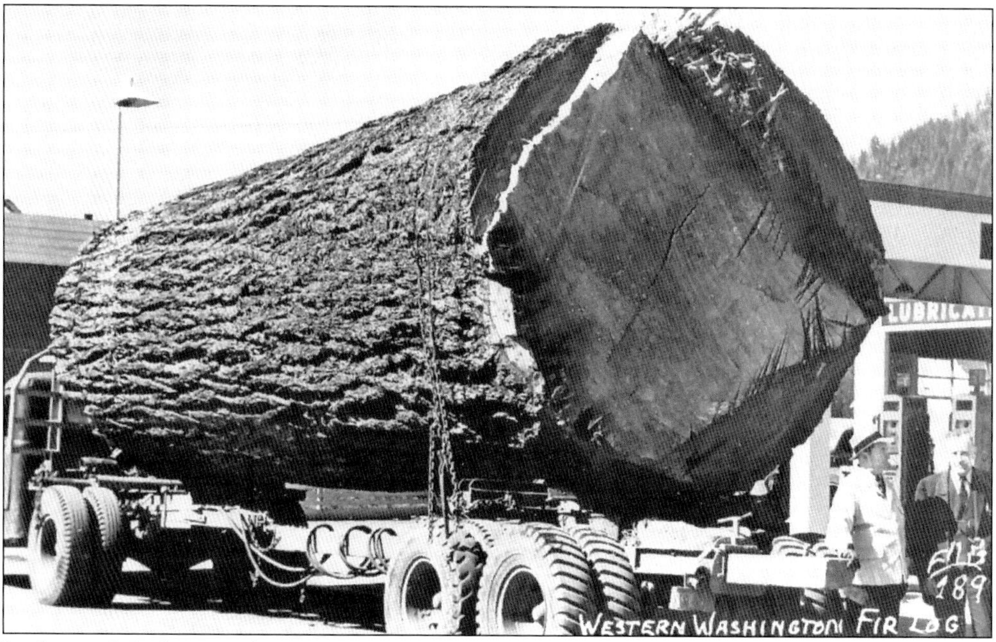

In this c. 1962 photograph, a Rayonier truck loaded with a big fir stops at the Forks service station. The driver stopped for a cup of coffee before heading to the mill. (CHSGWC.)

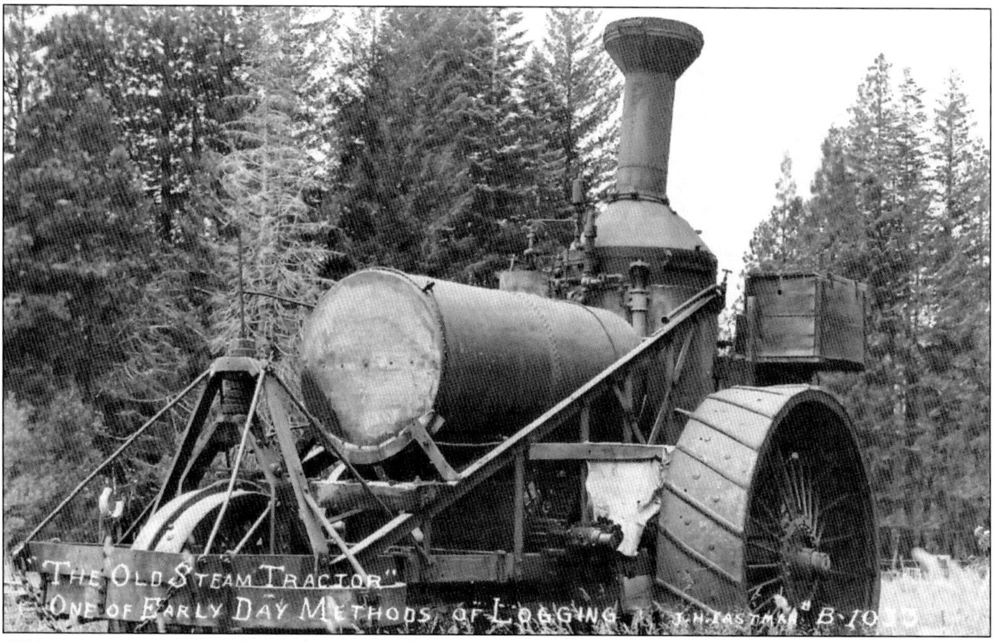

Old steam tractors, such as this one, photographed by J.H. Eastman, were in use in the early trucking days. (CHSGWC.)

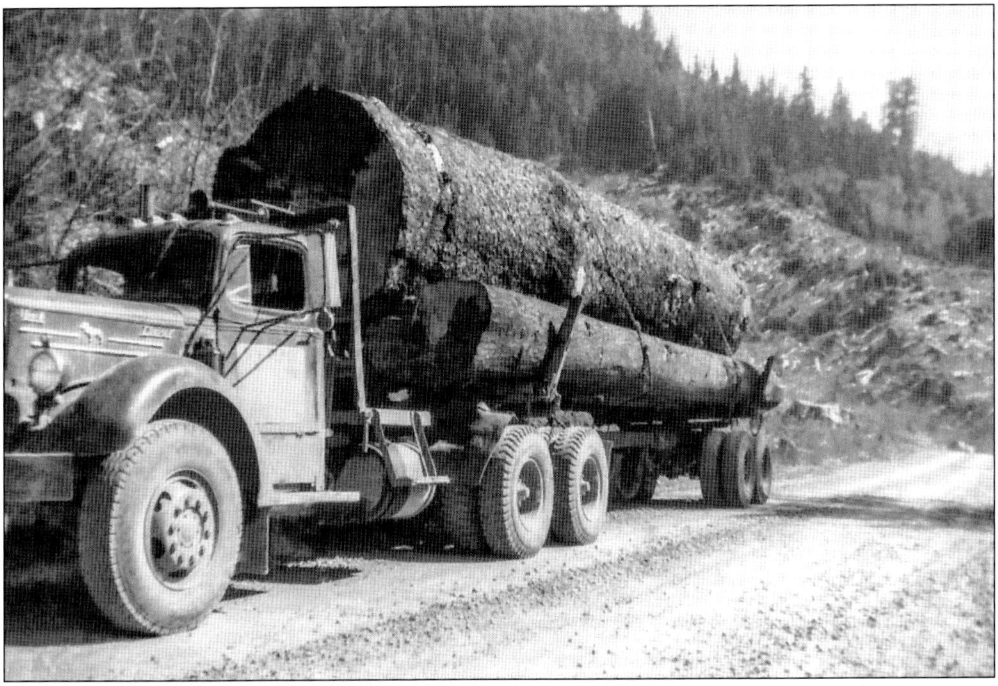

This big Mack truck, with its bulldog hood ornament winking in the sunlight, is hauling a big load out of the Clearwater area in 1963. Off-road trucks with wider bunks were used to haul logs to the re-load (transfer area). They were too wide and two heavy for public roads, having the capacity of a single railcar. (Gene Woodwick.)

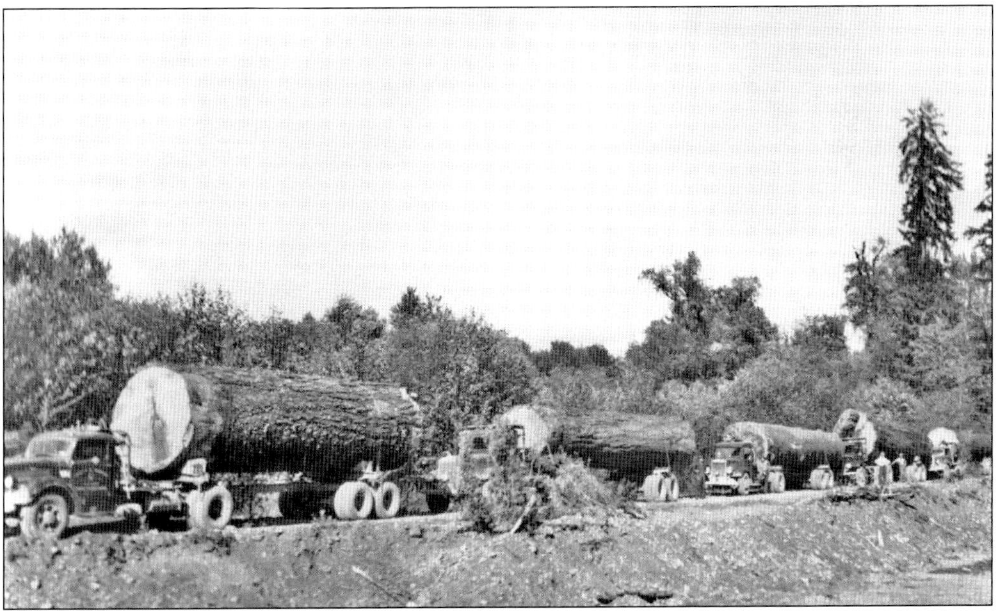

These five truckloads, from just one tree, must have been something to see. This Les Morgan photograph was taken around 1954. (AMH.)

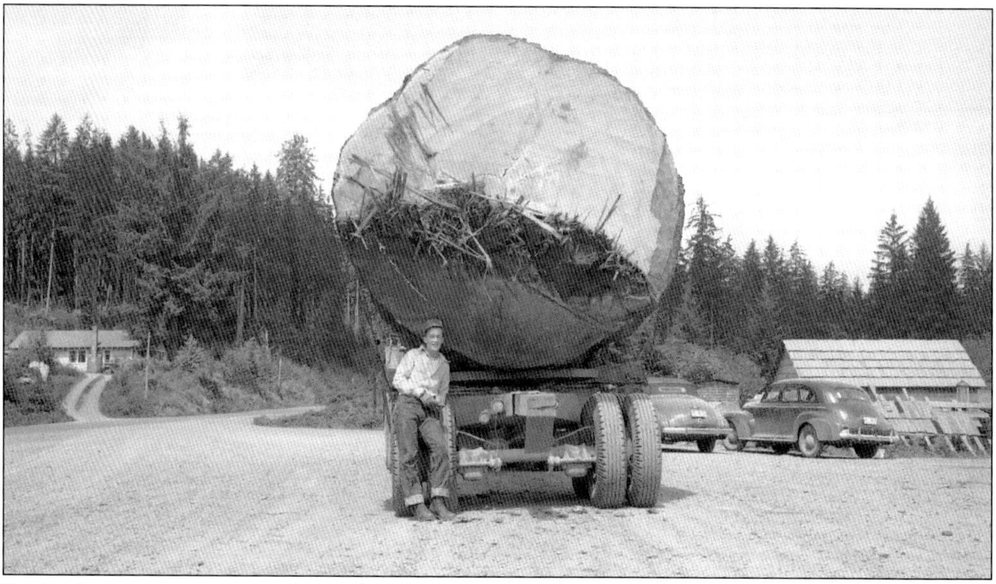

Big loads were a common sight in the 1960s, as at this loggers' café on SR 109 and the Montsano cutoff road. (PM.)

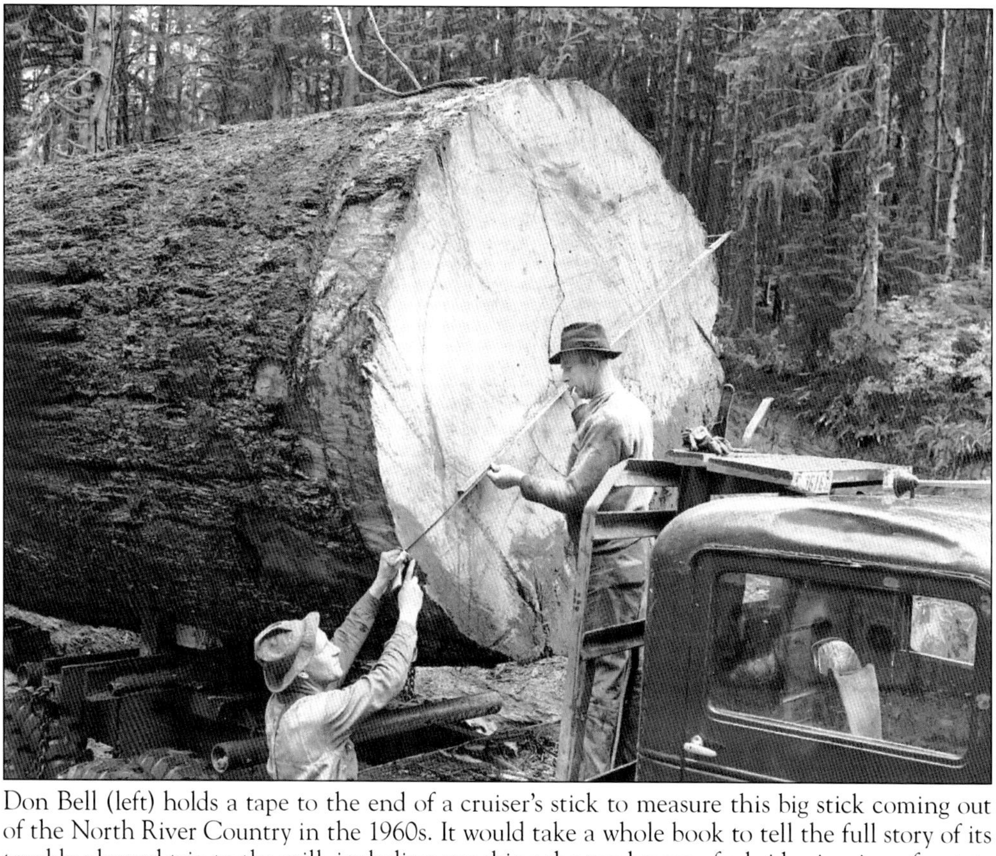

Don Bell (left) holds a tape to the end of a cruiser's stick to measure this big stick coming out of the North River Country in the 1960s. It would take a whole book to tell the full story of its trouble-plagued trip to the mill, including smacking the top braces of a bridge in view of a state patrolman. (PM.)

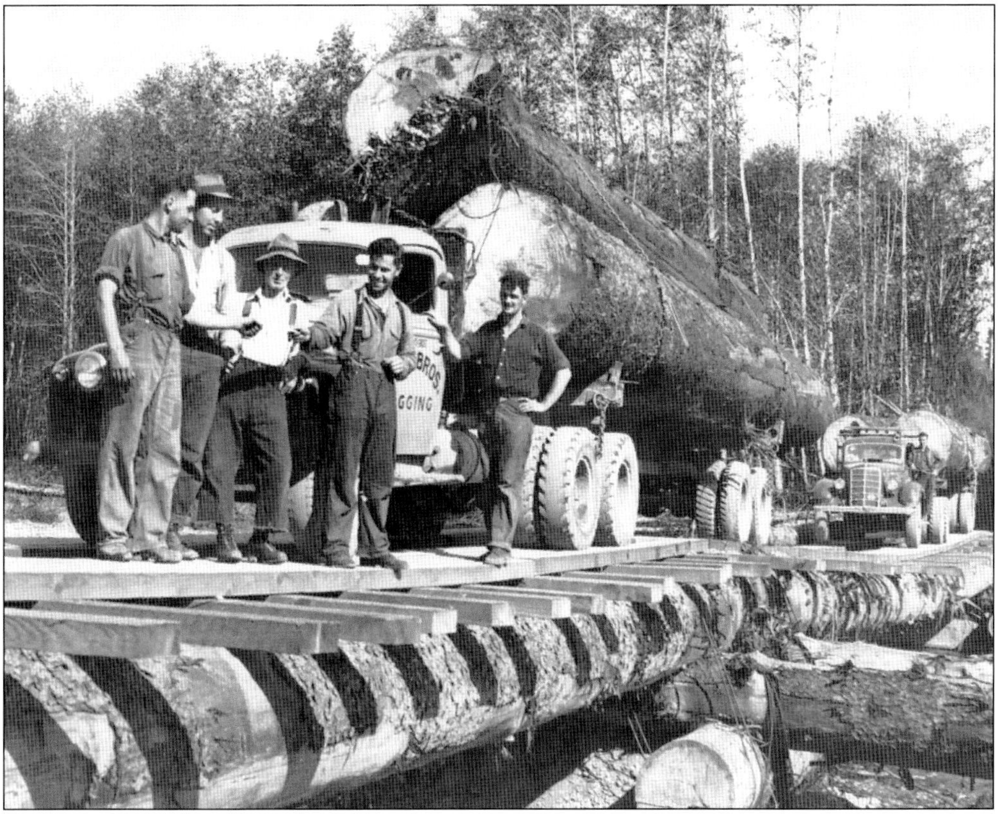

Mayr Brothers began logging in North County on the Carlson claim on the North Shore of Lake Quinault in 1938. The crews were working on the Clearwater in the 1960s. Here, one of the trucks is bringing in veneer peelers over the Queets River Bridge that Mayr built in 1943. (MNB.)

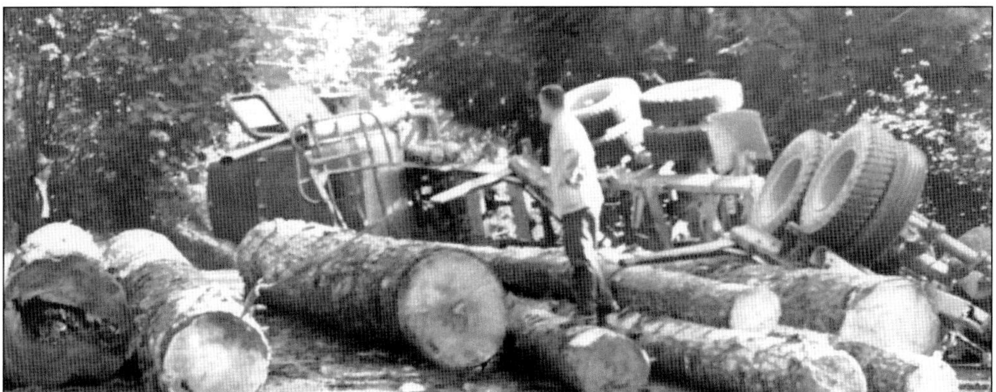

Not all trucks made it safely to the mill. This unhurt trucker was using some rather colorful language in 1964 as he looked over his truck, which wrecked when the chain binders broke and dumped his load. (Gene Woodwick.)

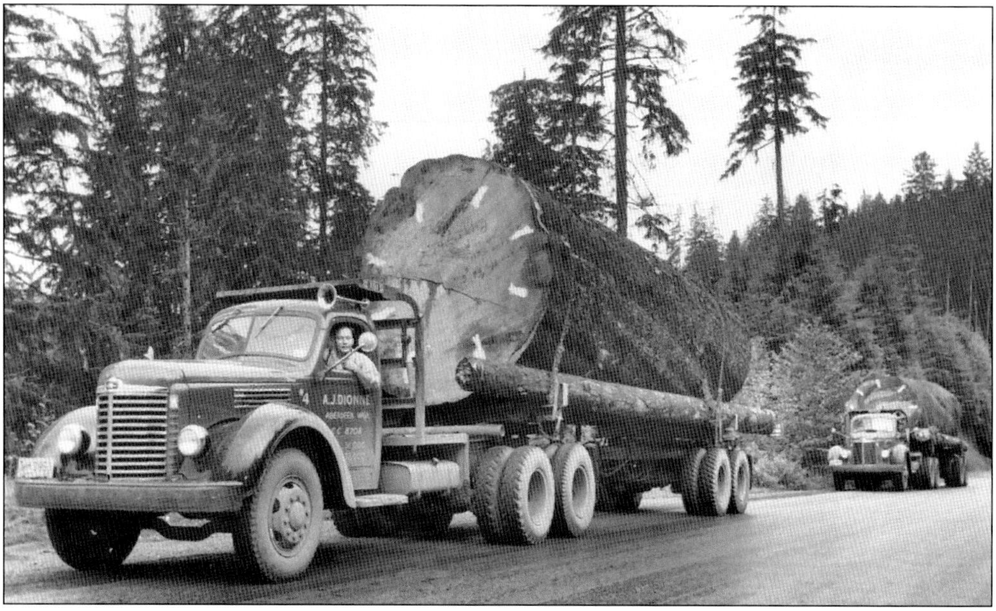

Big logs out of the North County area show white paint marks, for identification at the mill as Bureau of Indians Affairs (BIA) reservation logs. They are bunkered with logs and chains. (Gene Woodwick.)

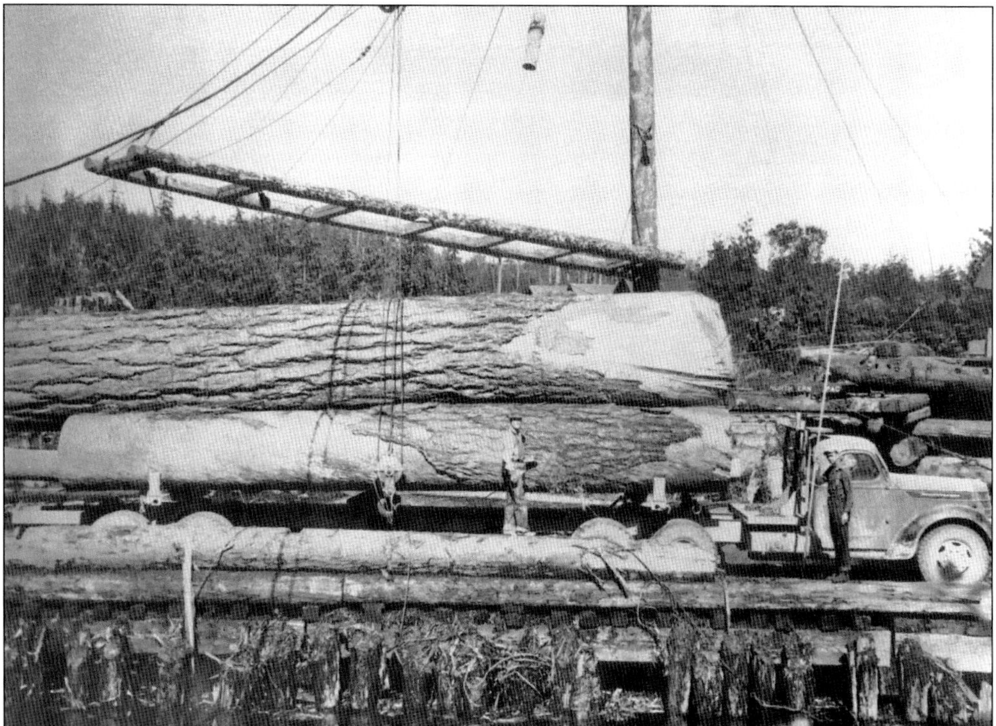

This was considered a fairly normal-sized load, seen here arriving at Aloha Mill around 1950. The sawmill, which produced 100,000 board feet a day, burned down in 1979. It was an end to one of the last active company towns, after 70 years of operations. (MNB.)

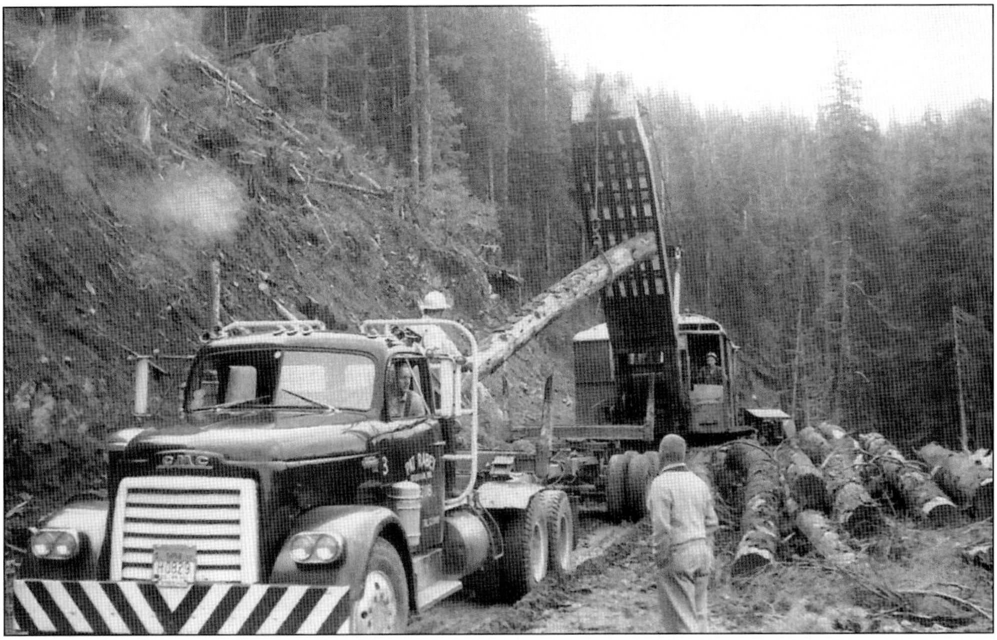

A Rabey Trucking Company landing crew uses its hayrack boom to pile a load on the Jimmy driven by David Rabey. The truck will then head out to Grays Harbor mills. (RTC.)

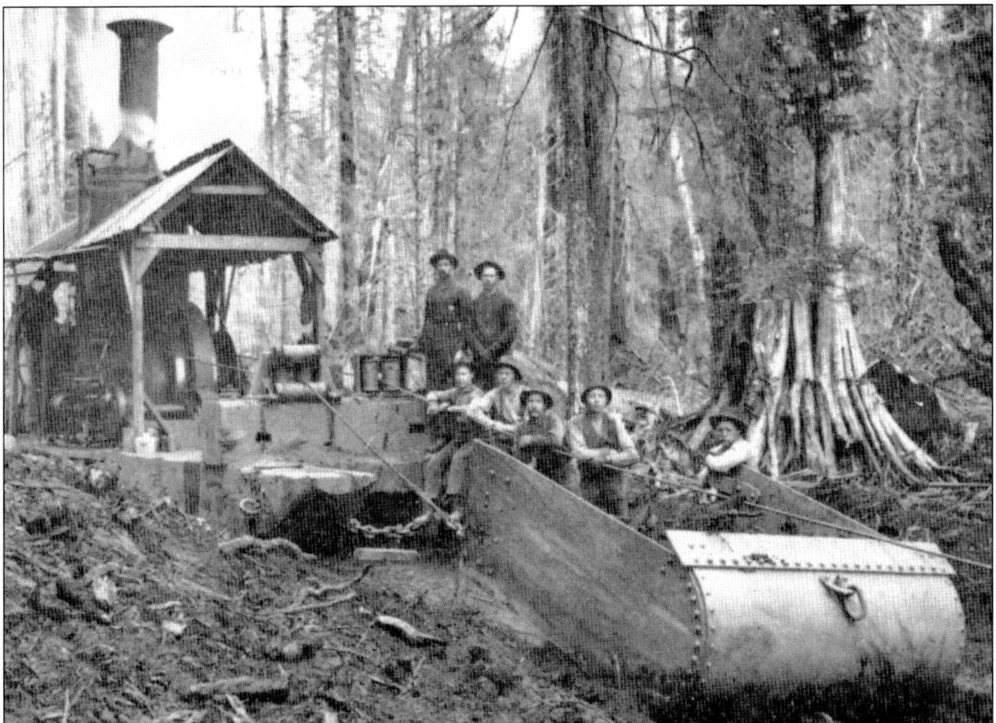

This early news photograph shows a Bagley Scraper, hauled around by horses and steam donkeys. The device, invaluable in early road building, dragged men and hand tools to a site while smoothing the way for a pole and/or a skid road. The Bagley was the precursor to the bulldozer. (CHSGWC.)

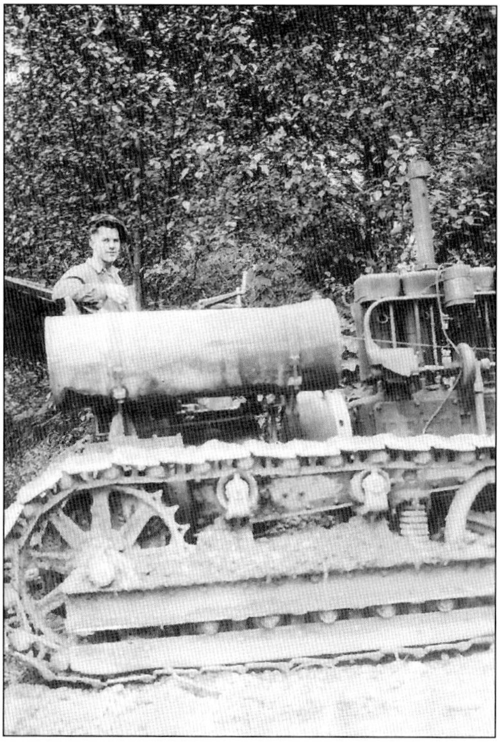

The first crawler tractor to enter the Pacific Northwest was the Schafer Brothers' small Cletrace. The "Cat" quickly became a staple of nearly every logging outfit after its appearance in 1921, following a merger of the Best and Holt companies, which became the Caterpillar Company. The Allis-Chalmers Cat came out in 1935. (CHSGWC.)

This Cat is using a bummer arch loader to pull a massive spruce log into the loading area. (PM.)

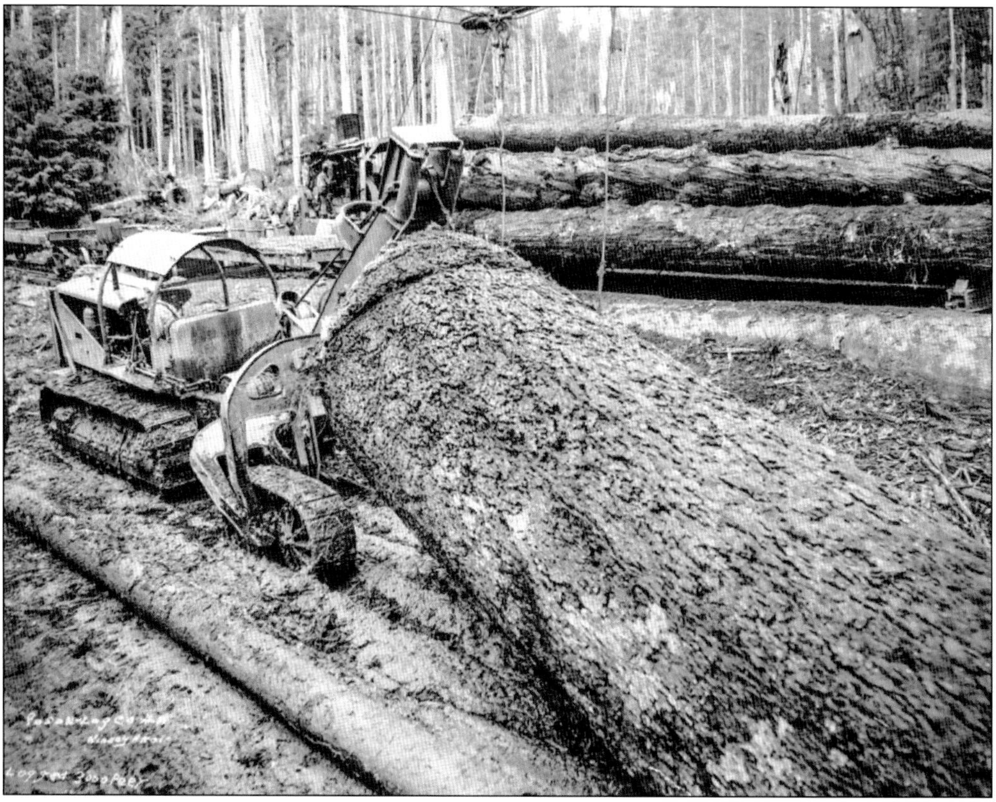

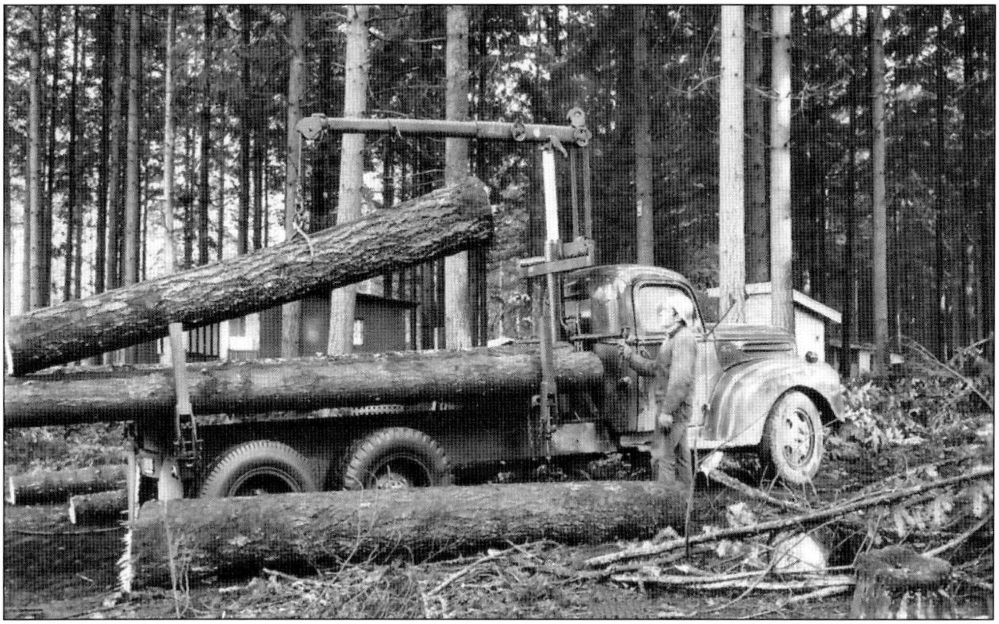

This early self-loading truck was a boon to the gyppo loggers. Self-loaders became the standard throughout the industry. Now, two men were able to fall, limb, cut to marketable lengths, and load logs and then head for the market. (CHSGWC.)

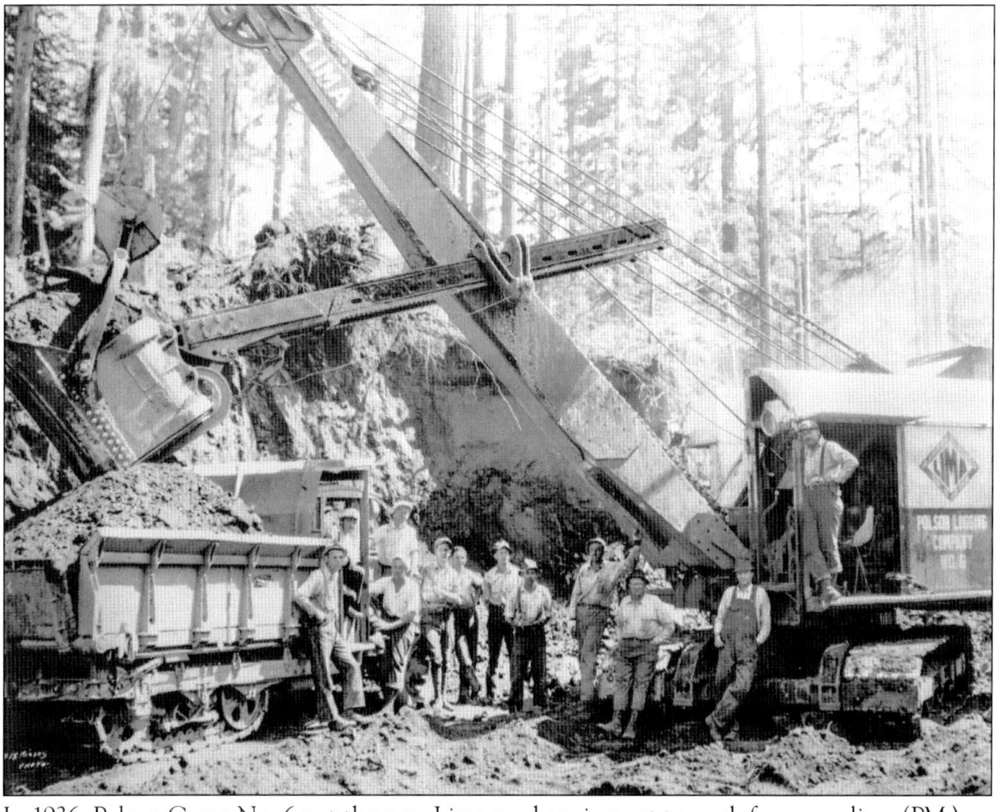

In 1936, Polson Camp No. 6 put the new Lima road equipment to work for a new line. (PM.)

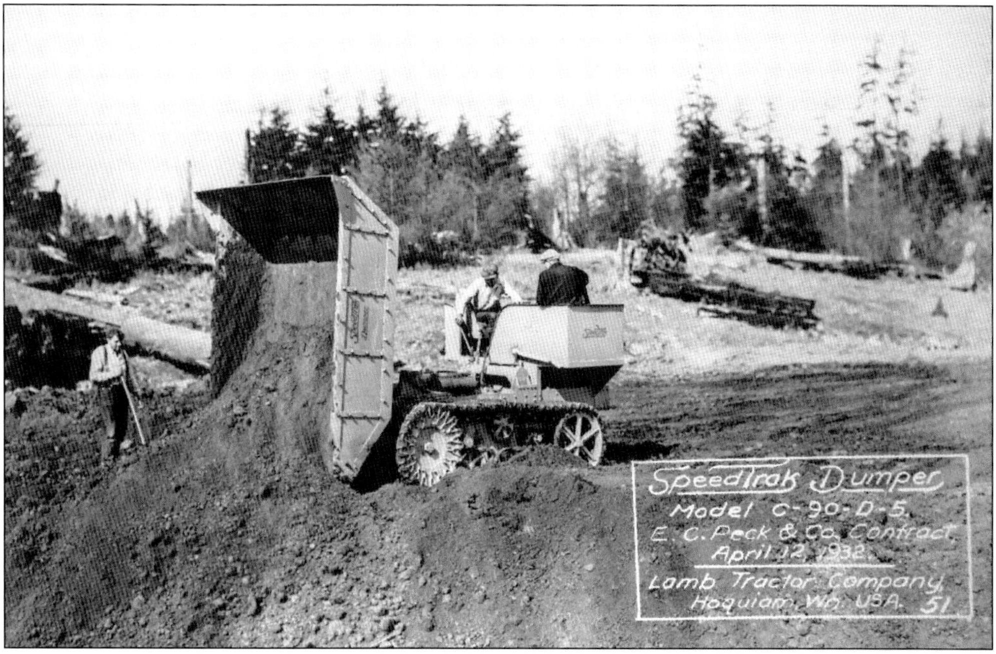

Lamb's SpeedTrak dumpster, invented in 1932, is said to have been the model for the World War II tank treads at Ruben Fleet's McDonald-Douglas airplane plant during California war production. (PM.)

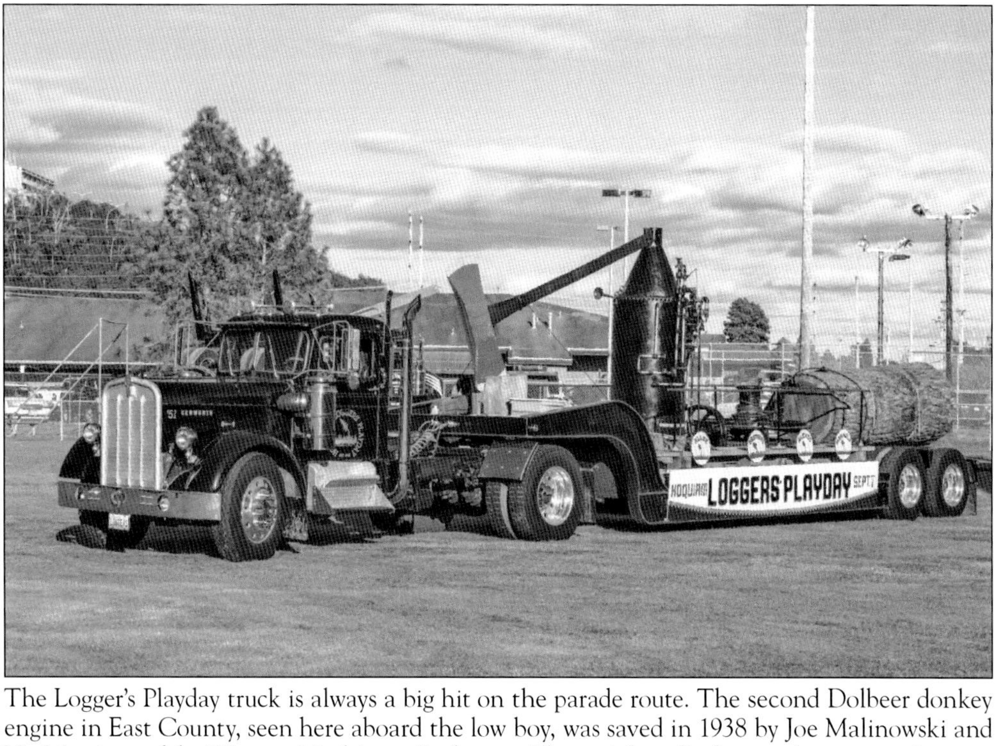

The Logger's Playday truck is always a big hit on the parade route. The second Dolbeer donkey engine in East County, seen here aboard the low boy, was saved in 1938 by Joe Malinowski and Vic Morrison of the Western Machinery Exchange. The upright cylinder ran the gear mechanism on the upright spool, much like the latter-day gypsy. (Brian Woodwick.)

Nine

Shippers and Shipping

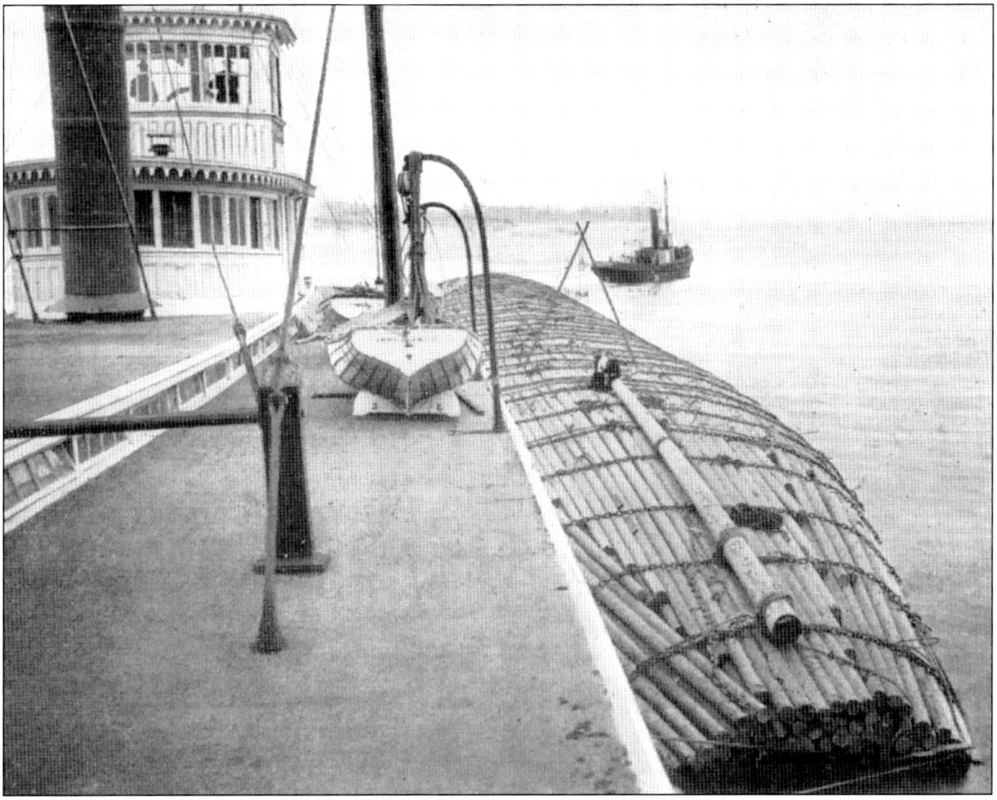

Using oceangoing rafts was an efficient way to ship large amounts of logs. The Benson cigar-shaped raft was 56 feet wide and 800 feet long. It had a 26-foot draft, held from 3.5 million to 4.5 million board feet of logs, and carried 13 feet above water. These rafts were towed 1,300 miles to San Diego from Grays Harbor. Davis rafts, 200 feet long, held 600 logs, or five to six million board feet. (AM.)

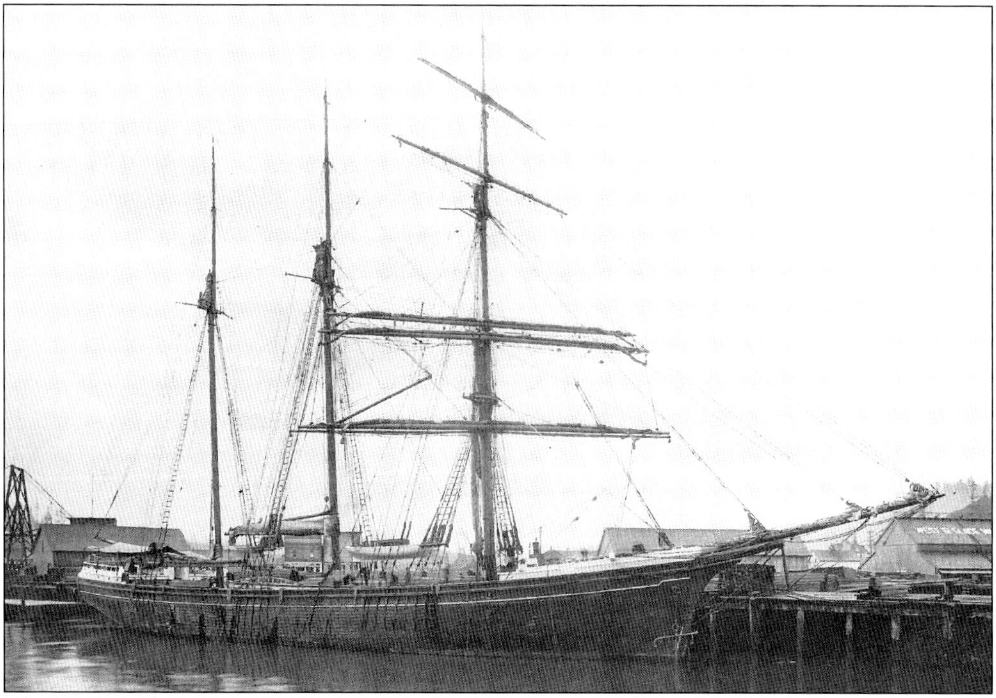

The lumber schooner *Benicio* was an American barkentine built in 1899. Here, it loads lumber at Slade Lumber Company on the Chehalis River in Aberdeen. She wrecked off the coast of Haiti in 1920. By 1929, a total of 45 vessels visited Montesano docks alone, loading 150 million board feet of lumber. (AMH.)

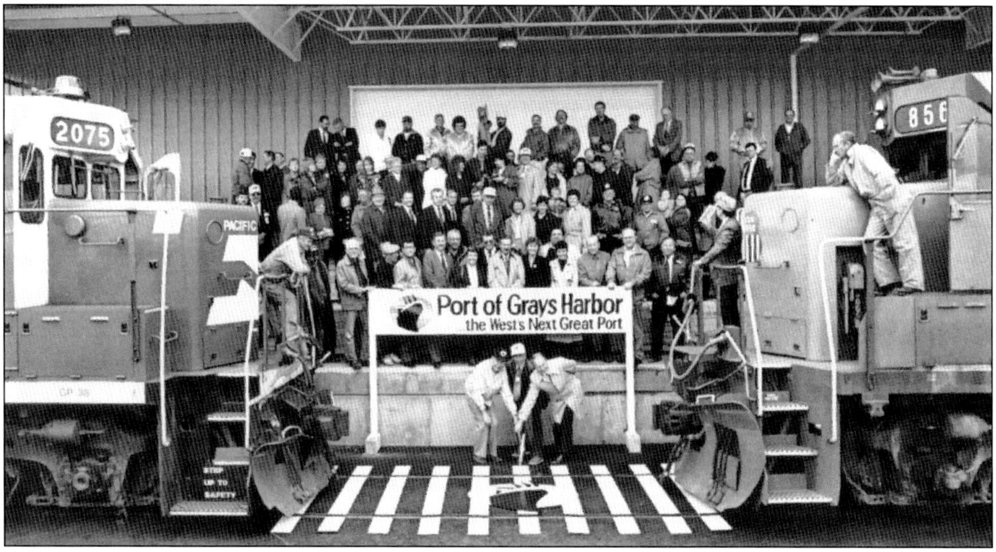

On the seven miles of tideland waterfront, 22 mills worked 24 hours a day. Their output for foreign markets necessitated the creation in 1924 of the Port of Grays Harbor, the second in the state. By January 1929, it was the largest shipping port in the world. It is still the only deepwater port on the West Coast. During the 1990 Marine Terminal Rail dedication, port commissioners Bun Lewis (left) and Russ Richardson (center) congratulate Cliff Mueller (right), executive director of the port, as they stand under the celebration sign. (CHSGWC.)

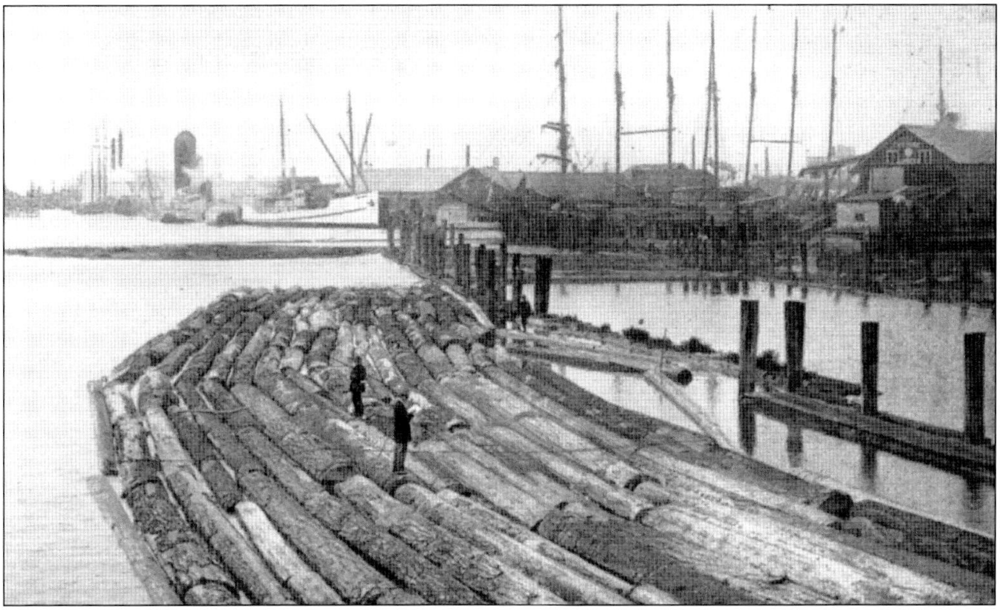

In the six decades since the first fir was harvested in the area, one billion board feet of timber was shipped out, making Grays Harbor the Lumber Capitol of the World. That amount of lumber, if laid end to end, would encircle the equator seven and a half times. (AMH.)

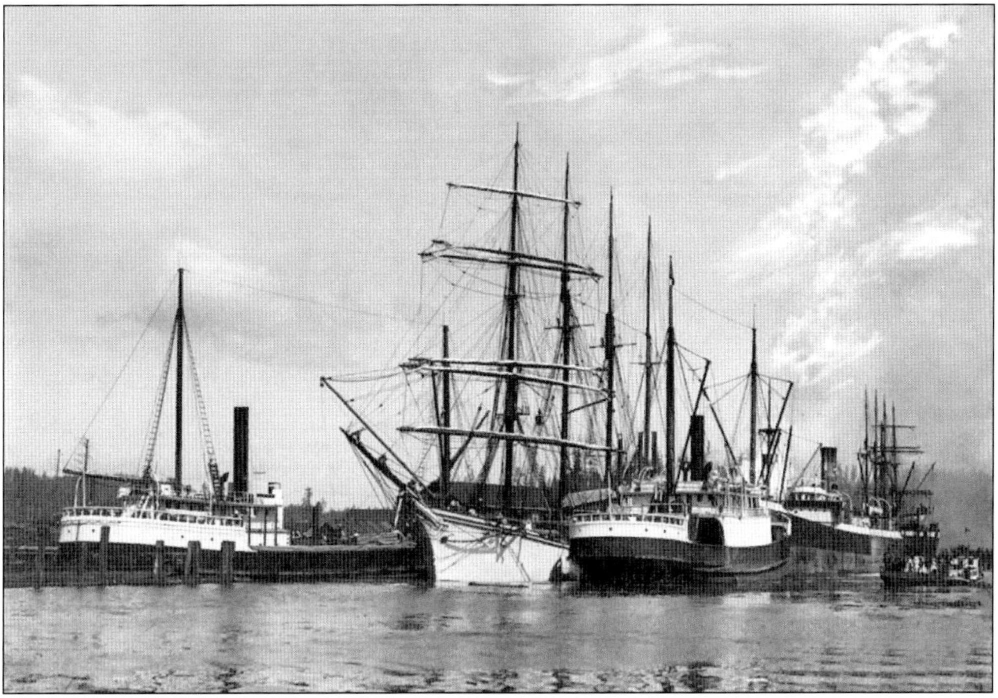

Hoquiam Lumber & Shingle Company, shown in this photograph donated by a Mrs. Soule, was built in 1902. Better known as the Lytle Mill, the company was originally owned by Robert and Joe Lytle. It was located upstream from the E.K. Wood Mill on the Hoquiam River. (AMH.)

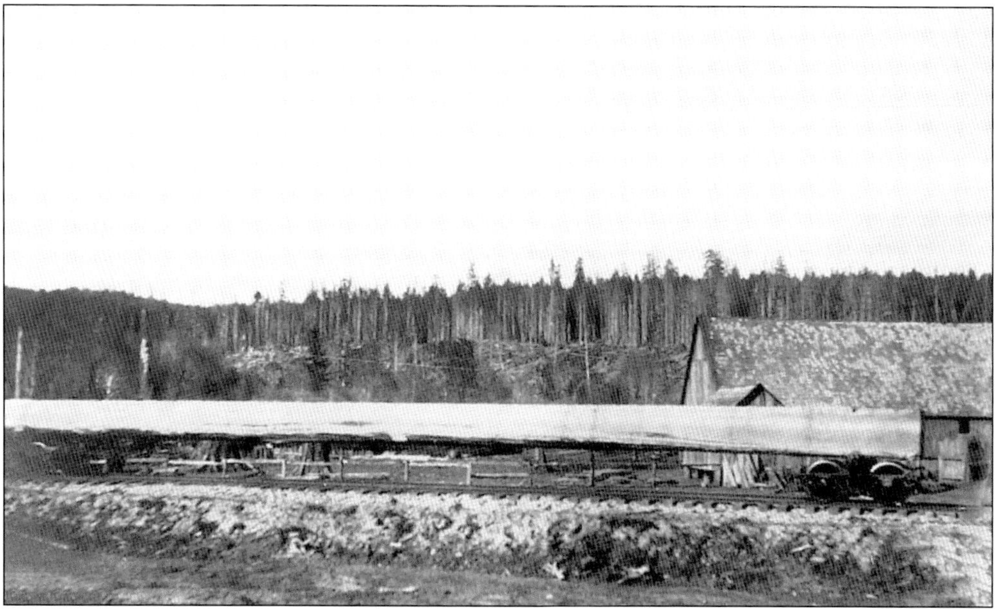

These massive timbers were on their way to Endreson Spar Company in South Aberdeen. The company was begun in 1902 by Lars Christian Endreson, a former shipbuilder. It produced many of the world's lumber schooner spars, including ones for the yacht of the emperor of Germany. (AMH.)

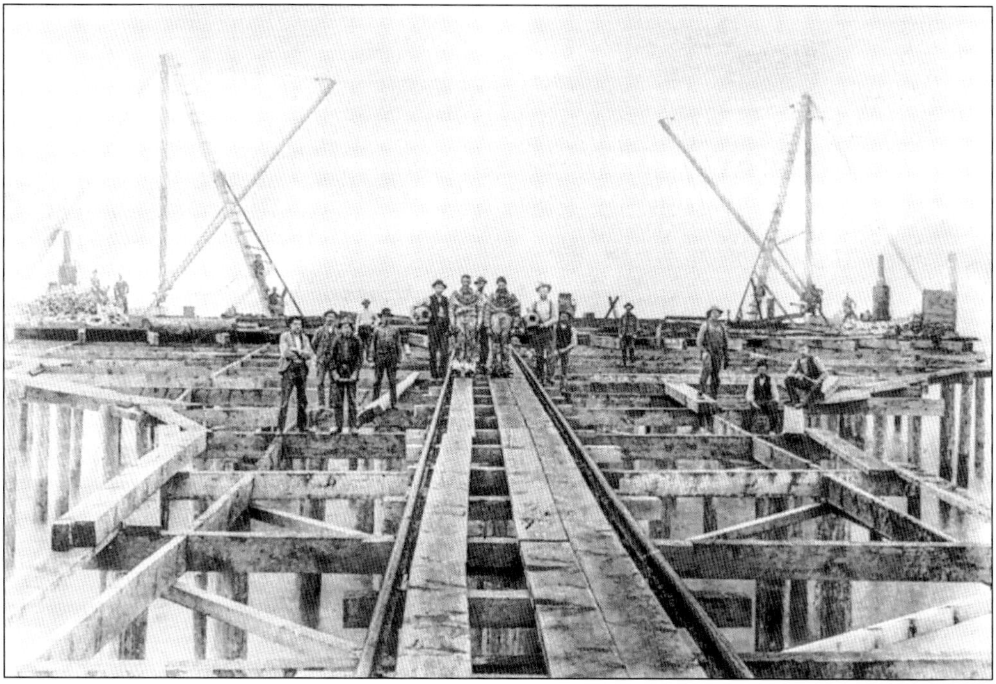

Grays Harbor City pier, located at the north end of Bowerman Basin, was built as part of the Walla Walla and Grays Harbor line from Centralia to Hoquiam. The line was part of the Hunt's railroad empire. Finances and outside pressure forced the company to cease operation, ending the dream of Grays Harbor City as the terminus to the Northwest. (AMH.)

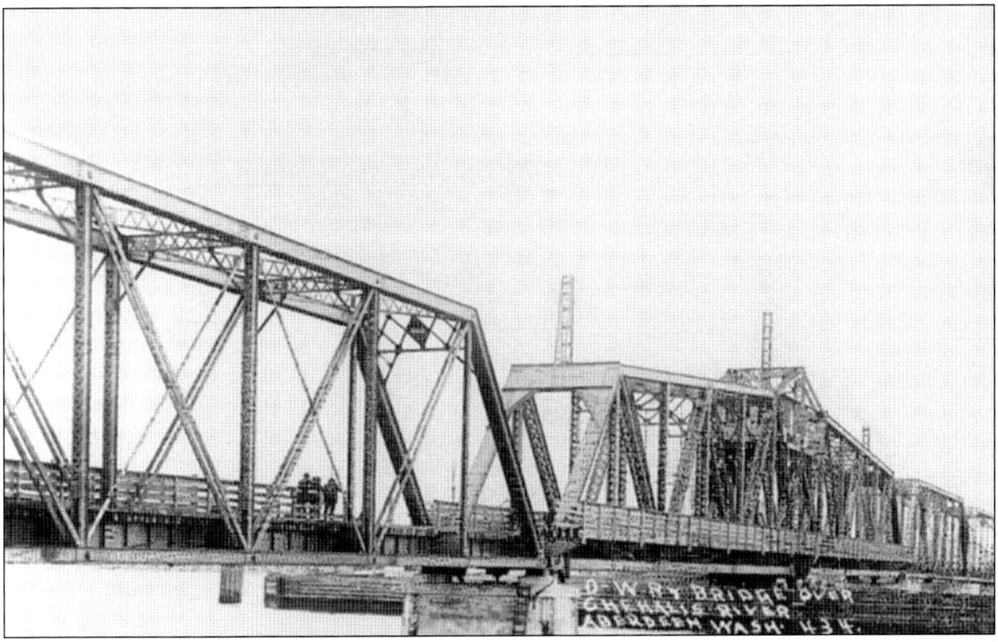

The first bridge over the Chehalis River in Aberdeen, built by the Oregon-Washington Railroad Bridge Company, was used to transport wood products to California and points east. (AMH.)

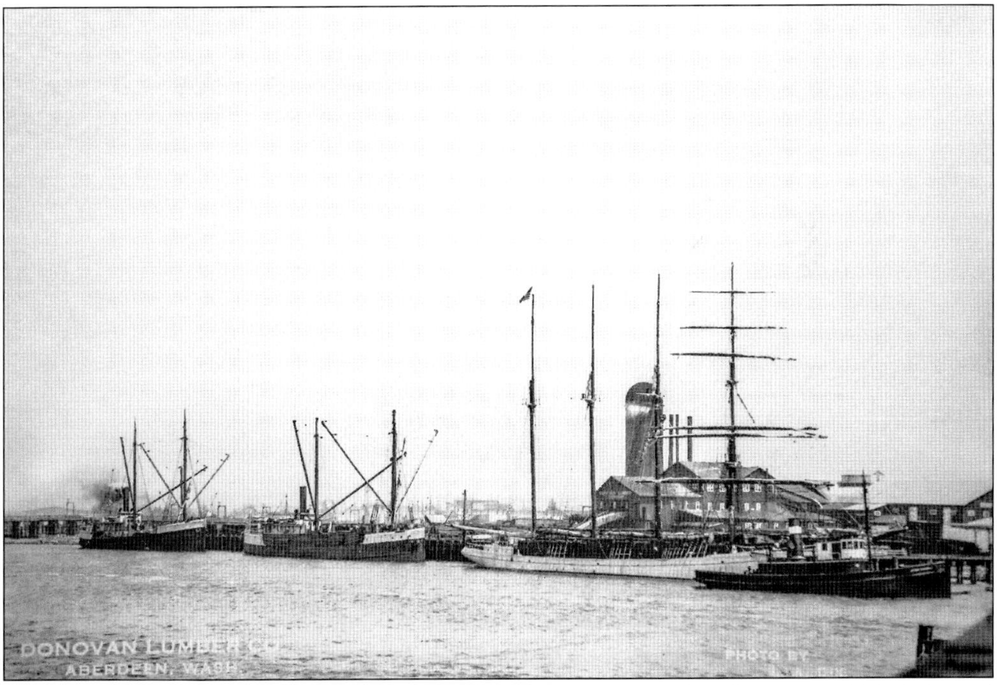

The lumber schooner *Blackheath* was a British bark. Here, it loads at Donovan's Mill 1, formed in 1910 by William Donovan Sr. and Patrick O'Conner. In July 1939, the mill was razed and the land was sold. When the Donovan Mill 2 closed on June 6, 1946, a total of 100 men were working there. (AMH.)

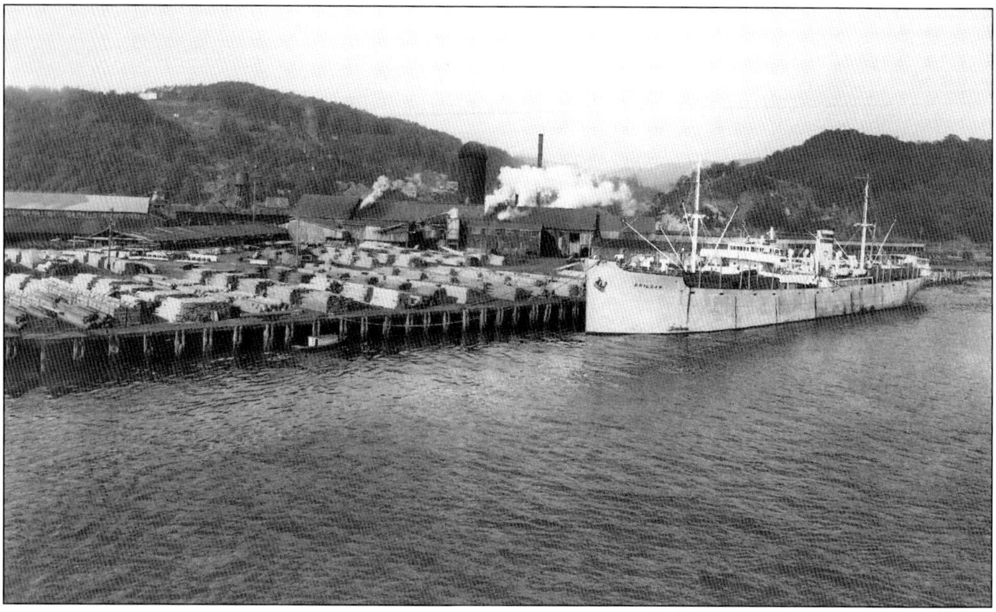

The SS *Chilar* loads lumber at the Wilson Brothers Mill, established in 1887. Think of Me Hill is in the background of this September 1930 Bliss B. Jones photograph. (AMH.)

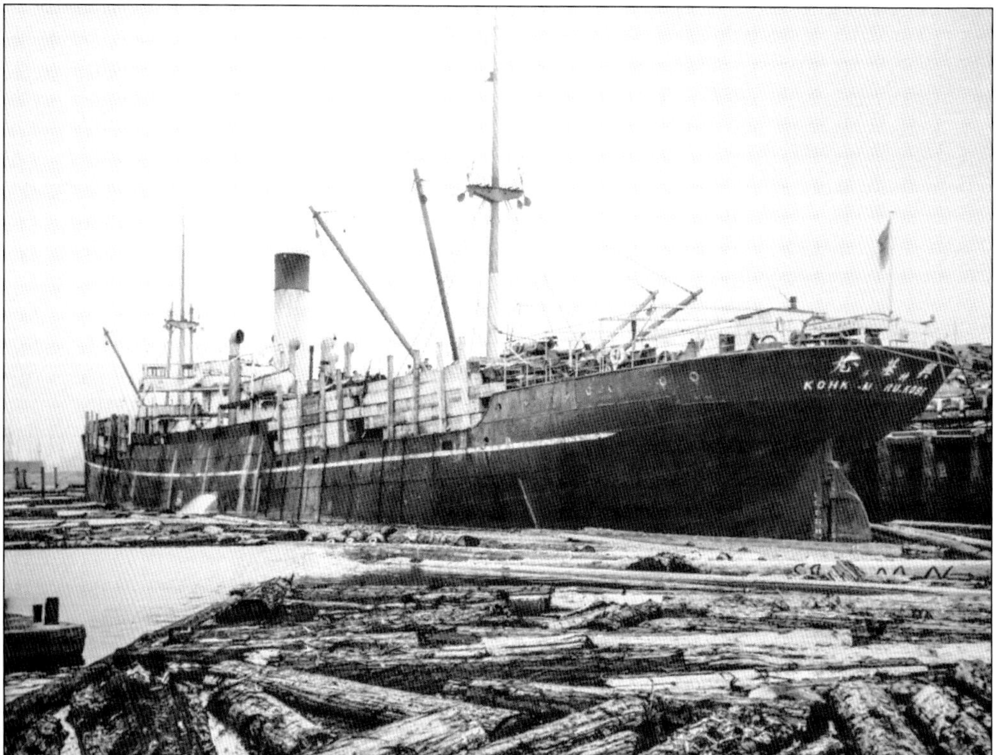

In October 1921, the Japanese shipments were breaking all records. Since January, 18 million board feet had been loaded. There were five Japanese steamers in port to take on five million board feet of lumber. (PM.)

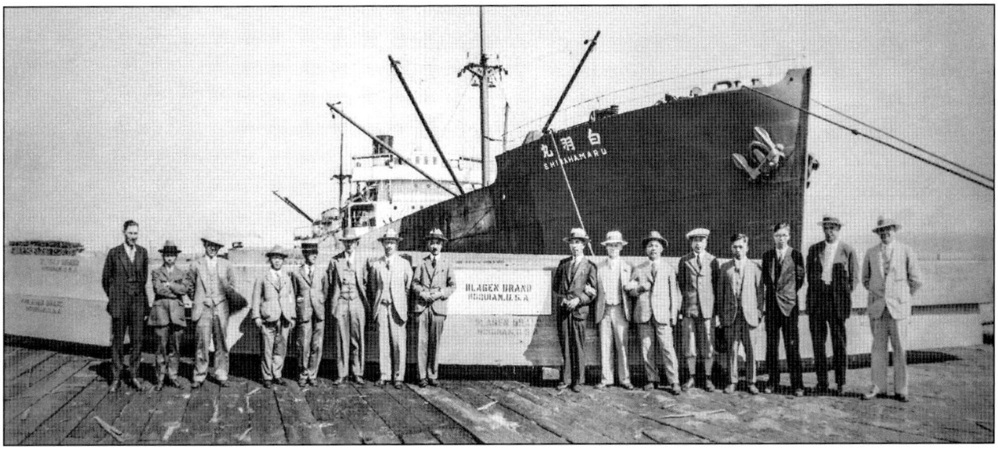

Port and Japanese officials celebrate the first load of timber products for the new Japanese market. (PM.)

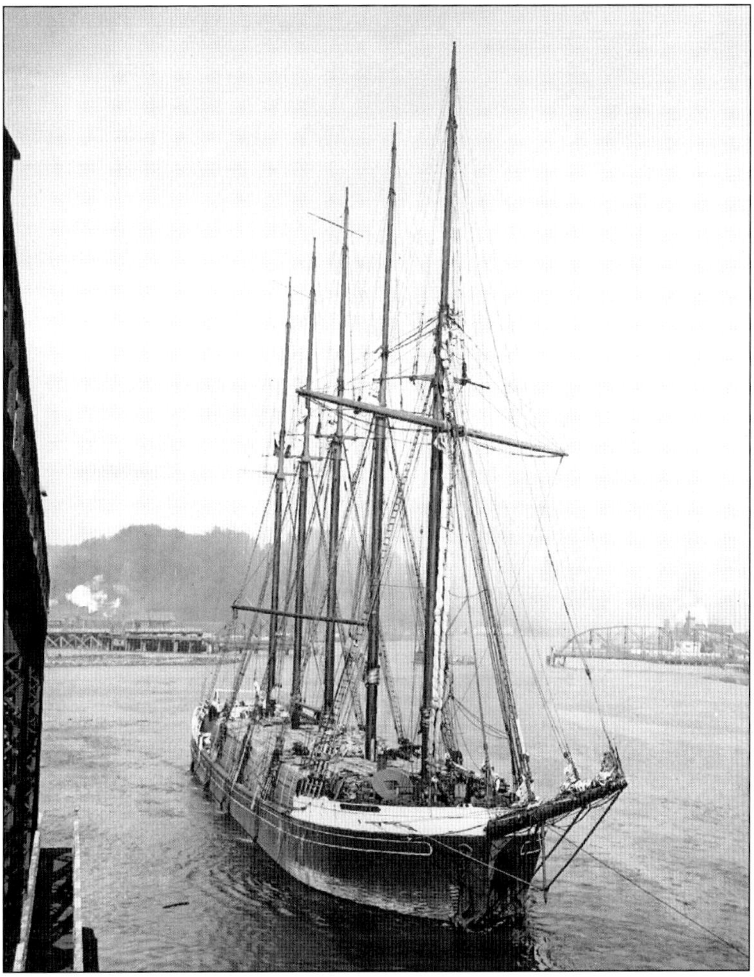

Capt. Ralph E. "Matt" Peasley captained the famous wooden lumber schooner *Vigilant*, which was built for the E.K. Wood Company and the North Western Lumber Company shipping to ports in the Northwest and South Pacific from 1889 to 1948. Peasley was made famous as "Cappy Ricks" by author Peter B. Kyne. The series of books, begun in 1915, were best sellers for over half a century. The *Vigilant* ended her days as the *City of Albernie* for the Canadian Transport Company. Here, Peasley heads out to sea from the Chehalis River with the old girl. (AMH.)

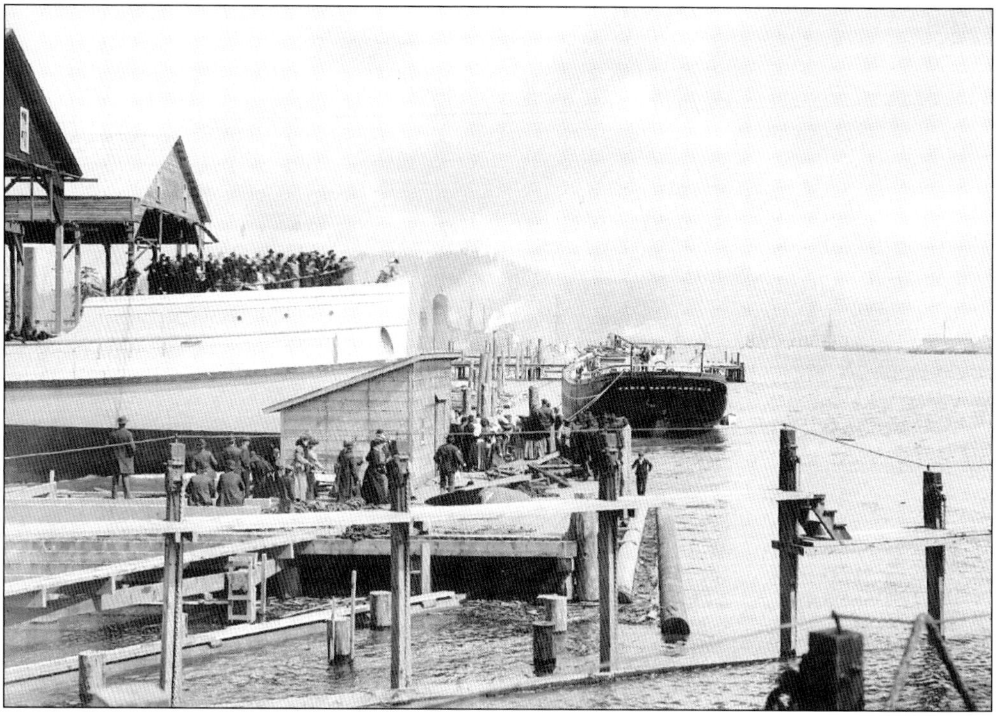

The demand for Grays Harbor lumber following the San Francisco Earthquake of 1906 was the impetus for the booming shipbuilding business. In 1906, the 582-ton "stemwinder" (steam schooner) *Quinault* was launched at John Lindstrom's shipyard in Aberdeen. Next to the pier is the *Sadie* from San Francisco. (AMH.)

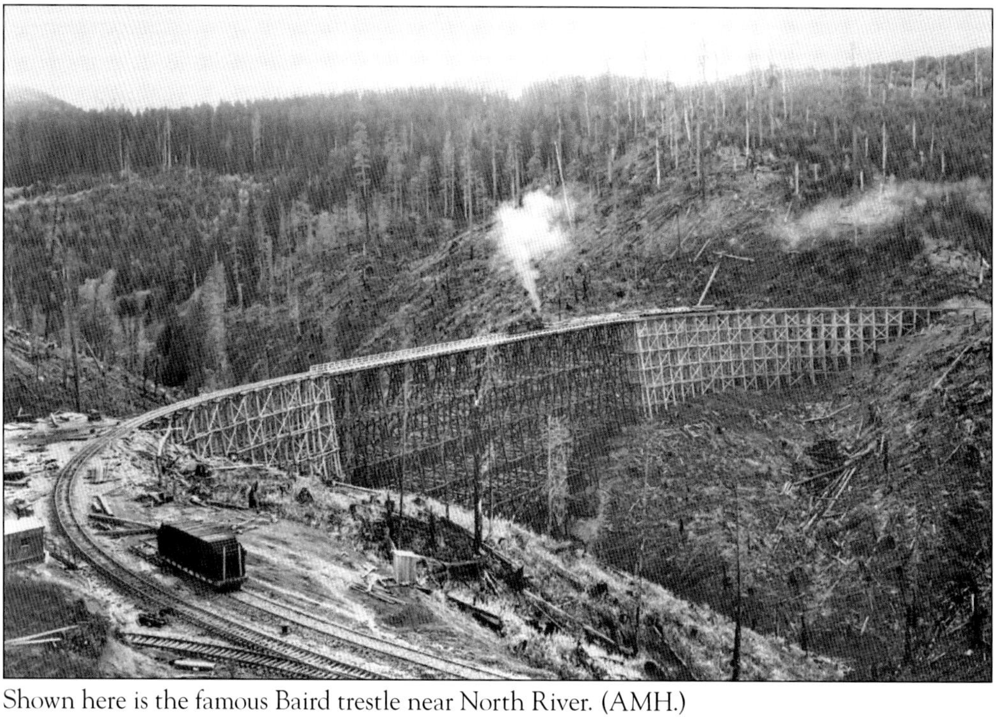

Shown here is the famous Baird trestle near North River. (AMH.)

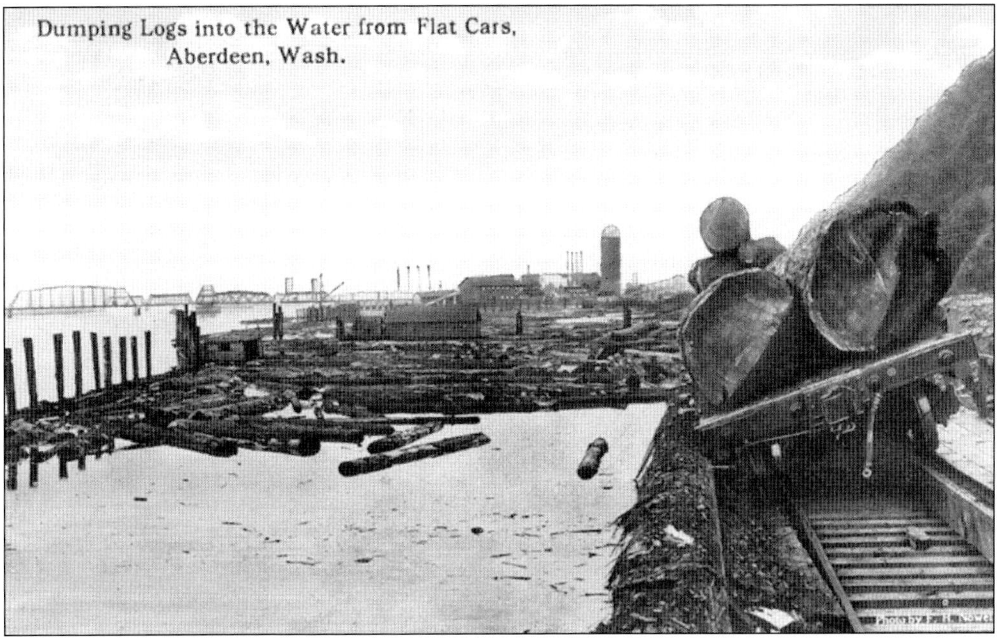

The rails carried flatcars of logs to the mills' booming grounds to await their conversion to lumber, which was then shipped via rail and sail. (AMH.)

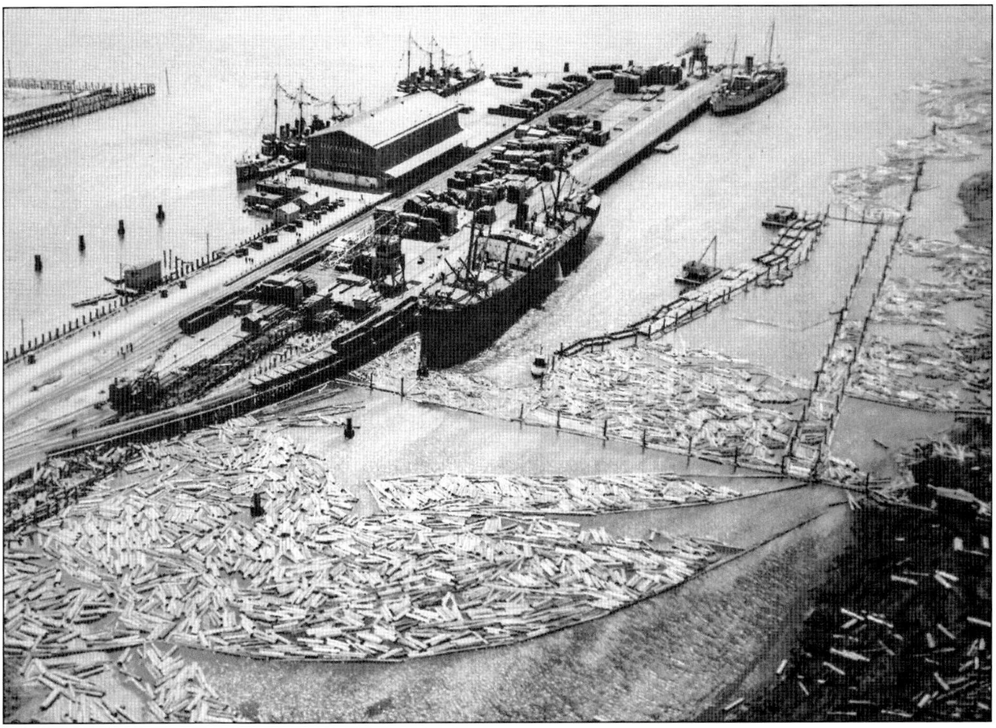

E.K. Wood, built by F.D. Arnold and James Karr in 1890 as the second mill in Hoquiam, shipped via a fleet of coasters, sail craft, and steamers to the offshore trade. The company ran seven schooners of its own. In 1899, some 126 steamers loaded at harbor mills, carrying away 49 million board feet of lumber. (PM.)

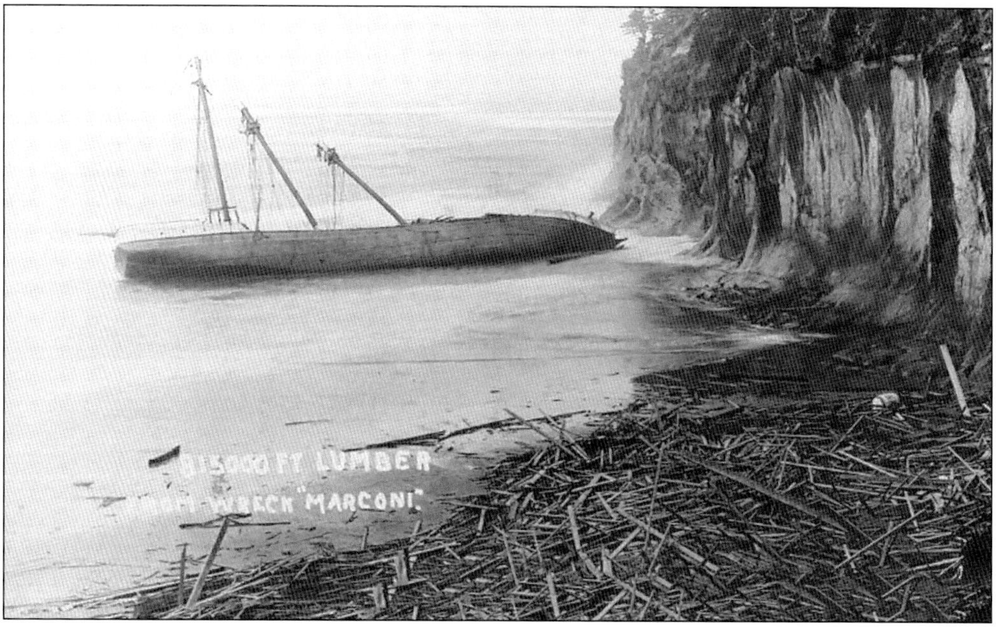

The four-masted schooner *Marconi* was being towed out of Coos Bay on March 23, 1909, laden with Grays Harbor lumber for Valparaiso. The hawser (towing rope) parted about midway over the Coos bar. She was driven into the south spit and was broken apart by the pounding surf. (AMH.)

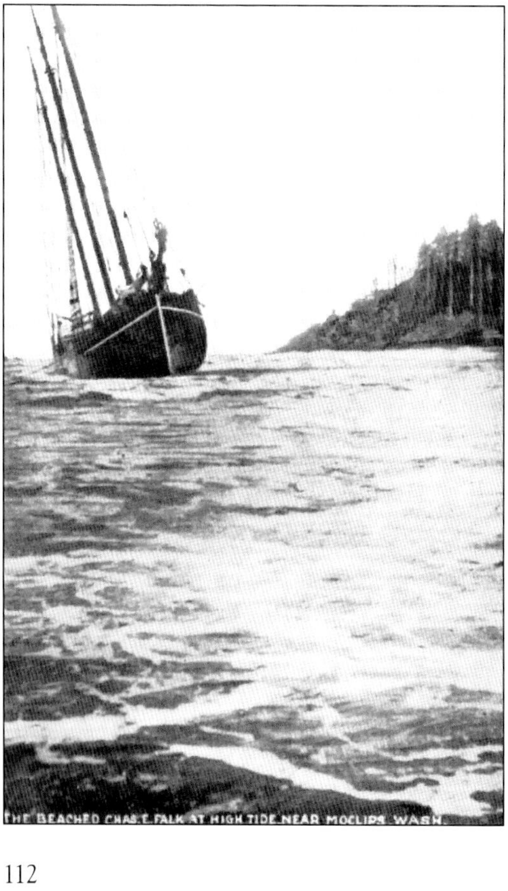

The lumber schooner *Charles E. Falk* beached on March 31, 1909, at high tide off the Copalis Rocks. The ship was unable to be saved. During the 1870s, the famous wooden sailing ships were diverted from New York and California's China tea trade routes to the Pacific Northwest lumber. (AMH.)

Ten

Heritage and Hereafter

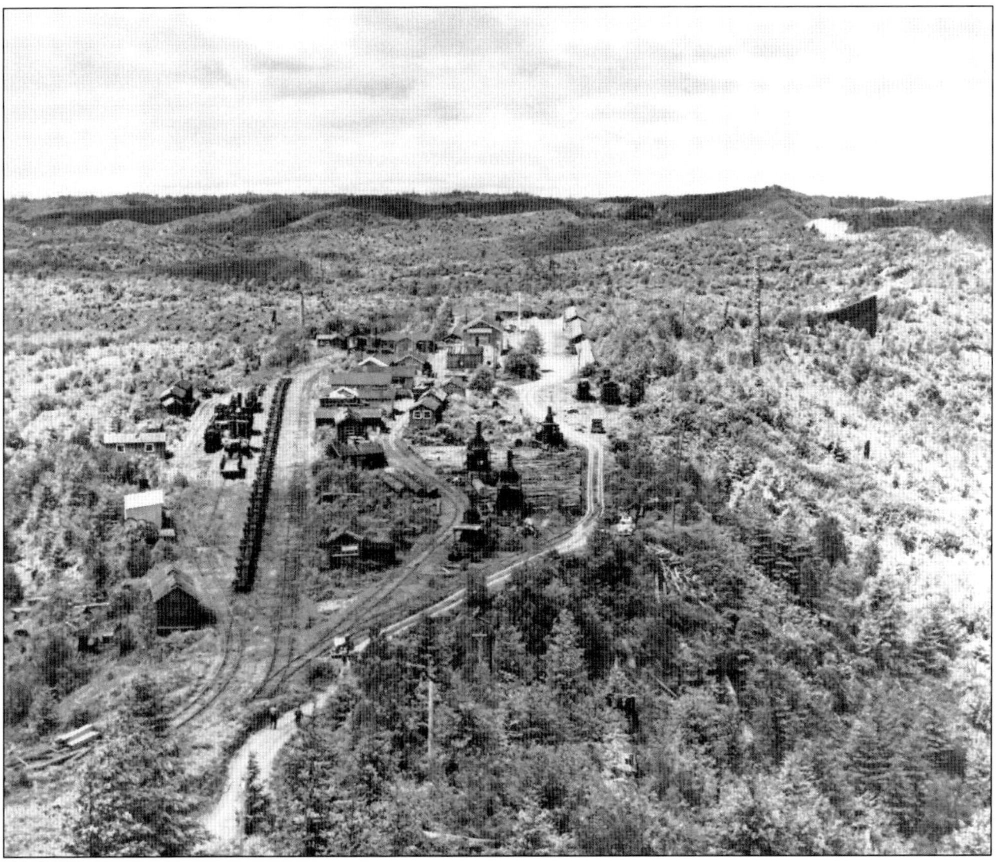

Clemons Company's Weikswood headquarters, southwest of Montesano, developed the first commercial tree farm, to "keep Washington green." (CHSGWC.)

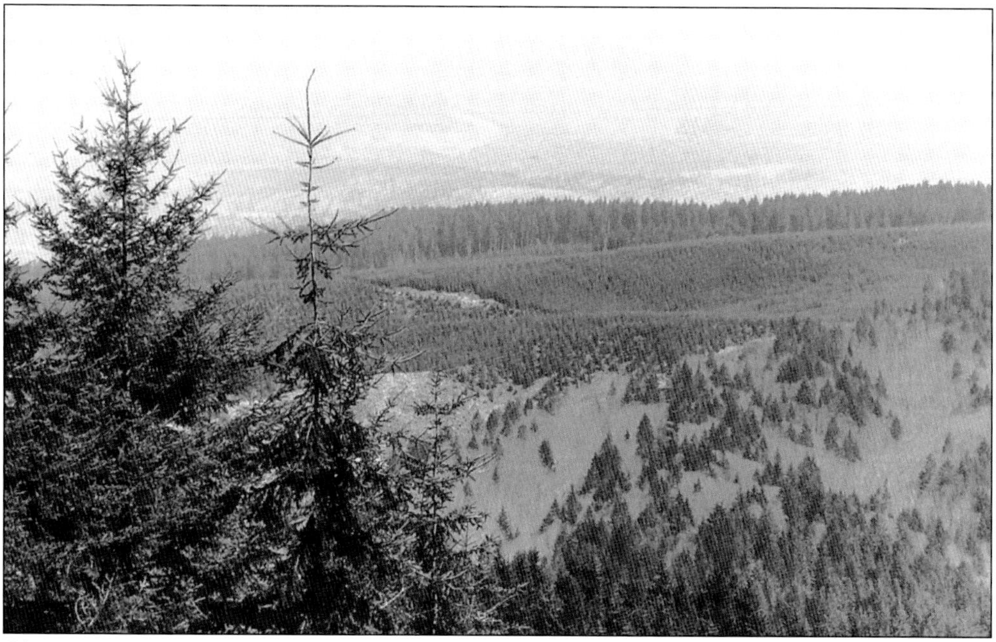

Clemons Tree Farm, operated by Weyco, began on May 1, 1941. The American Tree Farm system included members owning 93 million acres of timberland. This photograph shows a mixed stand of trees. (CHSGWC.)

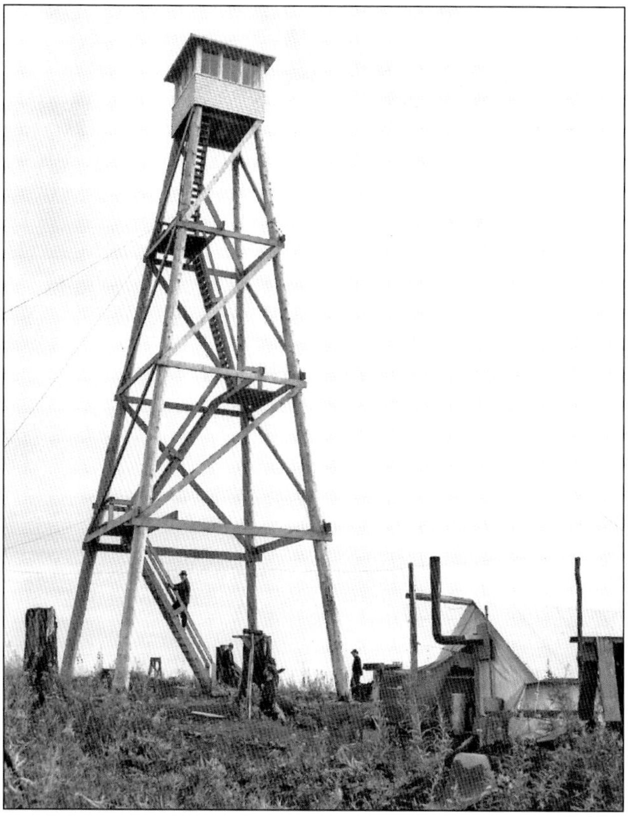

Byles lookout presided over Clemons Tree Farm, providing fire protection for thousands of acres of timber stretching from Montesano to North River. (CHSGWC.)

Fire-suppression communications were greatly improved in the 1950s with the proliferation of citizen band (CB) units, used by fire lookouts, truckers, and private individuals. This ranger is enjoying sunshine surrounded by a bed of irises in 1962 while monitoring calls. (CHSGWC.)

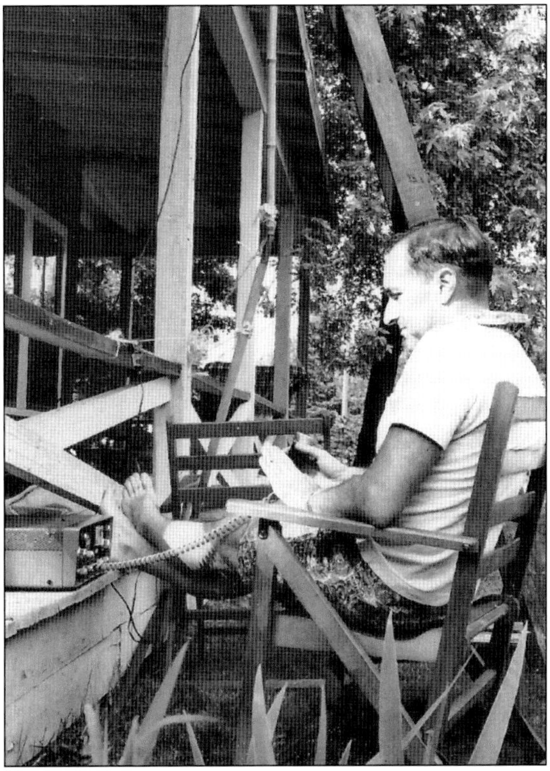

US Forest Service (USFS) employee Bob Ramstad hand-planted seedlings in 1940 at the Clemons Tree Farm. (CHSGWC.)

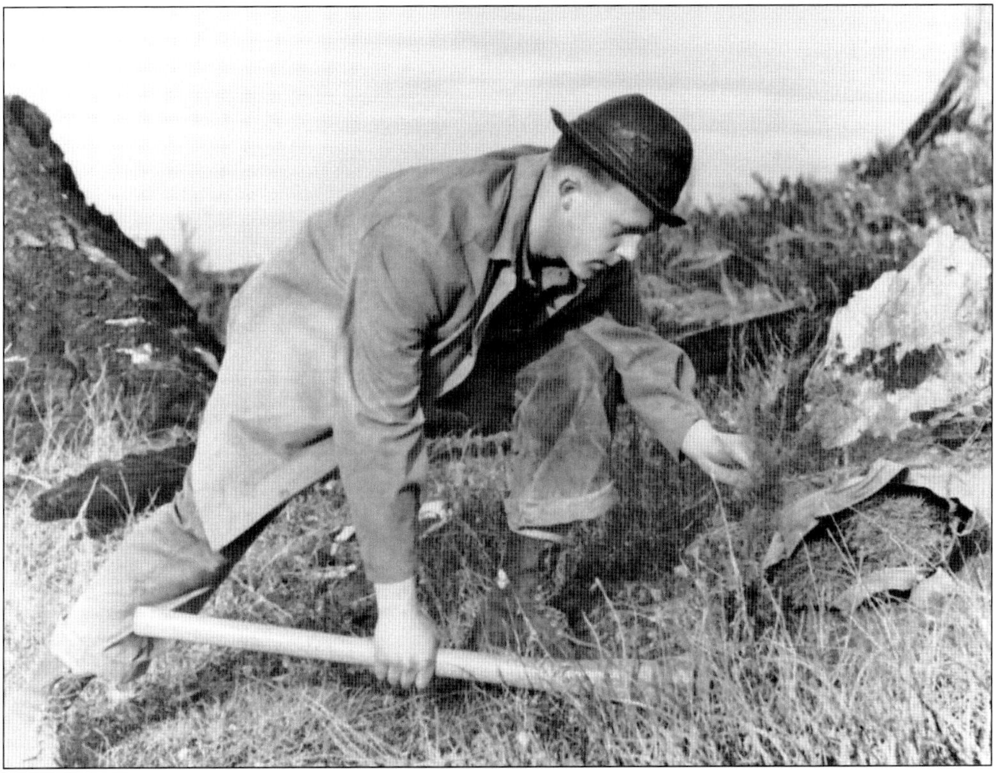

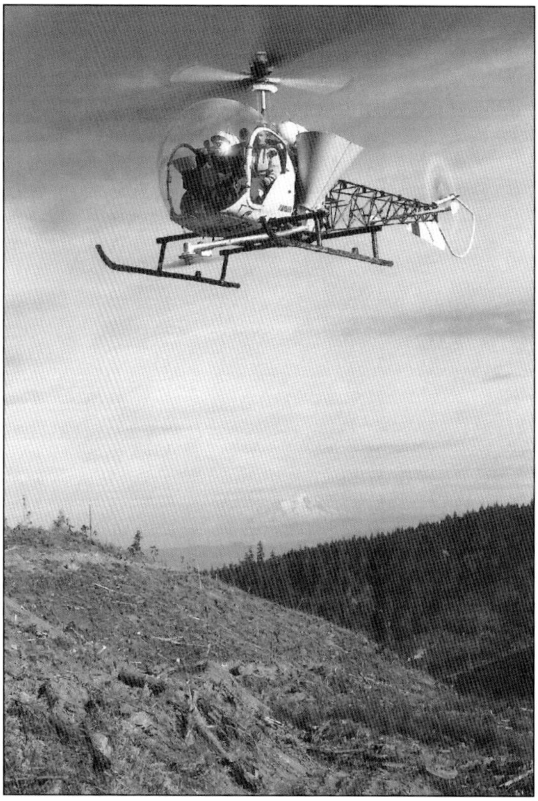

Aerial seeding in the late 1960s was not only efficient, it improved erosion damage in clear-cut areas. Returning Vietnam War veterans found many aspects of helicopter logging jobs a perfect fit for their wartime skills. (Weyerhaeuser Public Relations.)

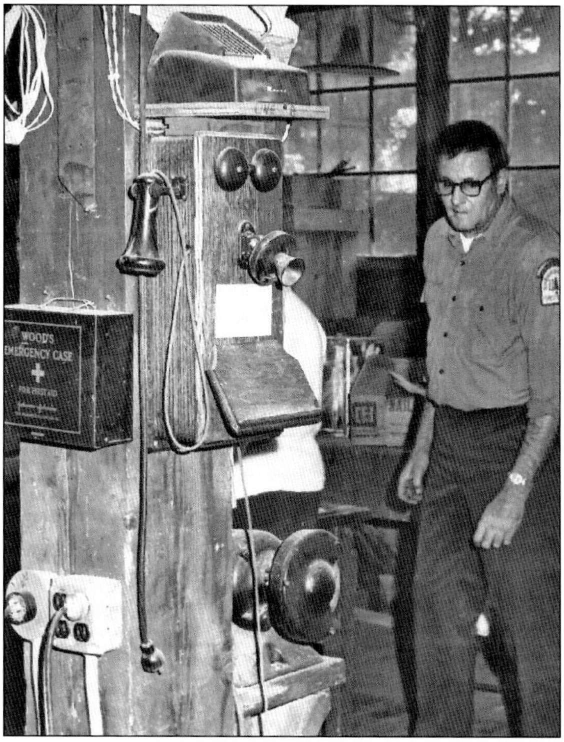

A USFS ranger checks out the communication area of a local ranger station around 1970. The old crank phone was being replaced with newer equipment. (CHSGWC.)

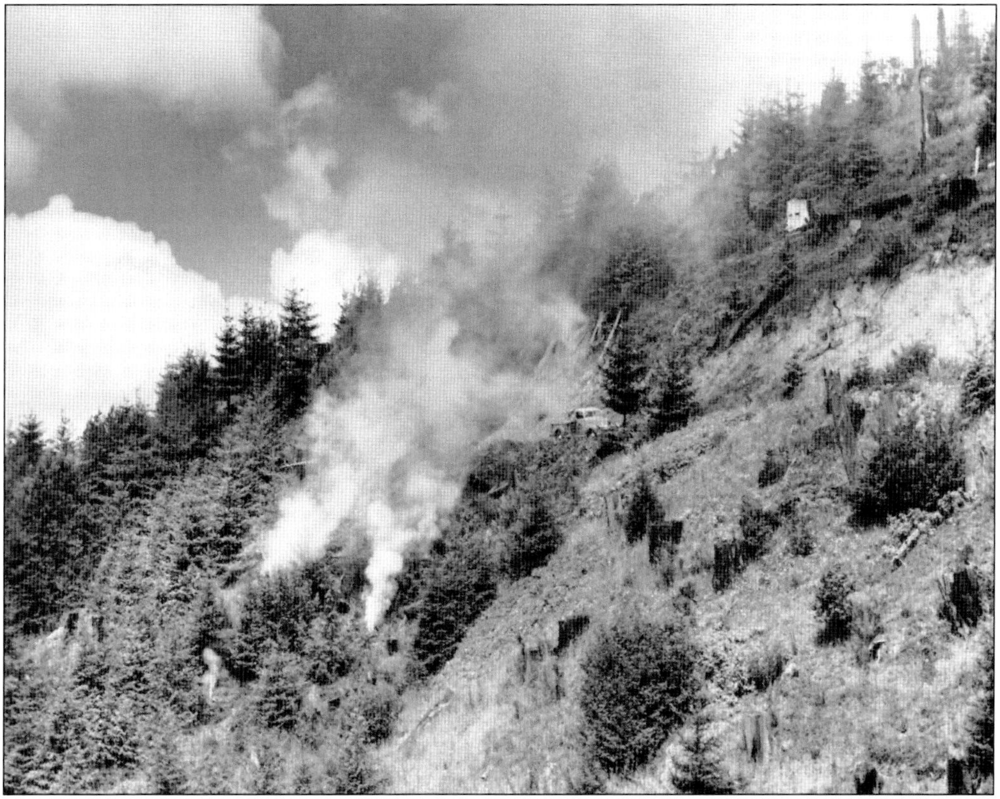

John Kagney staged a slash burn at the Clemmons Tree Farm in 1979 under USFS expertise. Slash burning not only served as fire protection, it returned minerals to the soil for replanting efforts. (CHSGWC.)

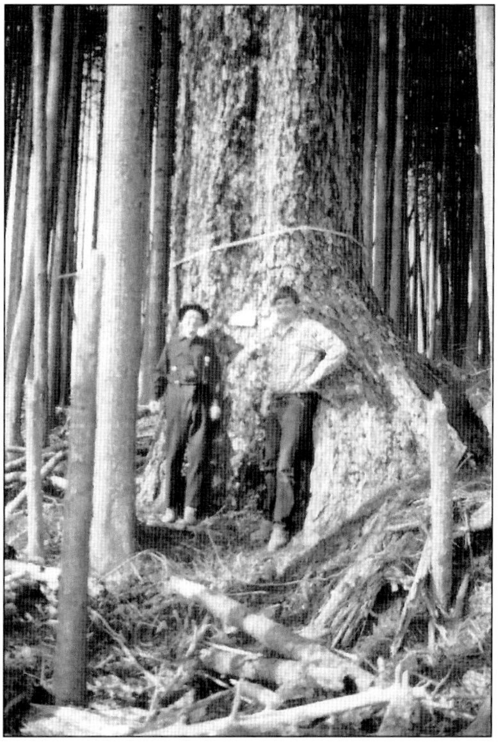

In October 1981, "Wes" Westby and his son Jim stand beneath a huge tree on Rayonier, Inc., land in the North Beach area. The tree saved Wes's life when he was lost in a storm while hunting. The company, renowned as a "people" corporation, responded to Wes's plea to preserve the tree. Today, it forms a minipark in the middle of a working forest. (Gene Woodwick.)

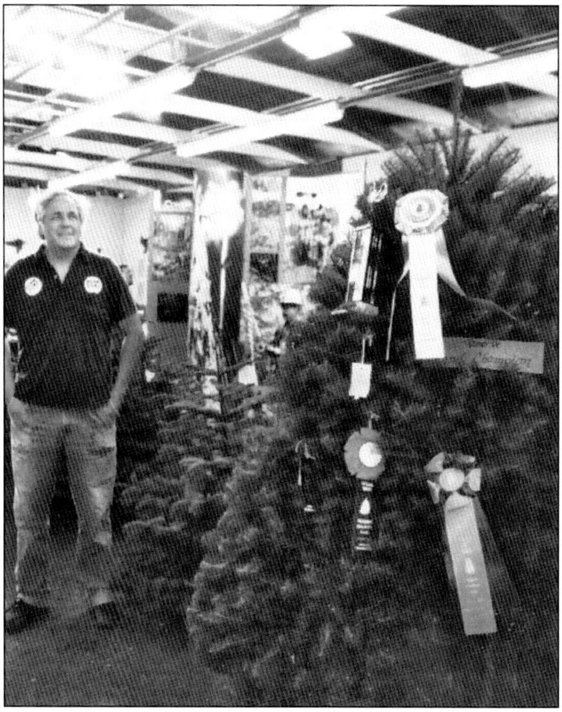

Larry Woodwick admires prize-winning trees from Grays Harbor Tree Farms. Blue-ribbon trees from the Hedlund Christmas Tree Farm in Satsop provided the official White House Christmas trees in 1998 and 2002. The Northwest Plantation Tree Farm furnished three White House trees in 2004. (Gene Woodwick.)

Big trees are the pride of Grays Harbor, which has supplied massive timbers for nearly every world's fair since the famous Chicago fair in 1893. This log was under construction to become the Aberdeen Chamber of Commerce office in the 1950s. Among products made by loggers to show off the big sticks were such things as a private railroad car, a vacation travel trailer, and sculptures around town symbolizing myths and legends. (AMH.)

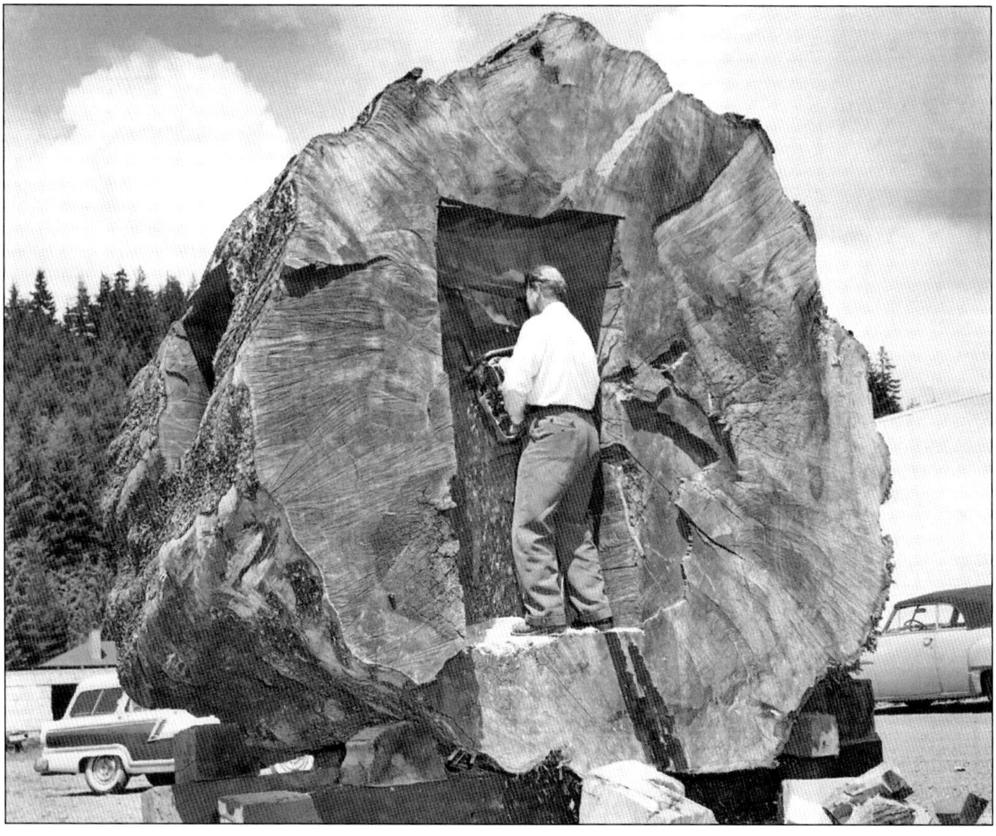

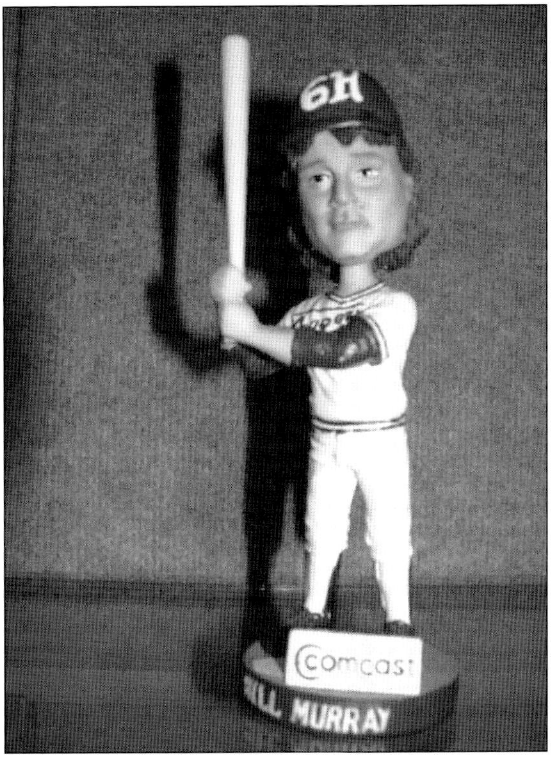

The 1977 Grays Harbor Loggers minor-league baseball team fielded actor Bill Murray in the summer of 1978. He was memorialized in this bobble head figure, owned by historian Roy Vataja. Too poor to buy uniforms, the players bought secondhand ones from a Utah team. They silkscreened their own names and numbers onto cloth and then sewed the patch over the former lettering. Note, however, on the sleeve a Golden Spike symbol remained. They just could not figure out a way to hide it. (Roy Vataja.)

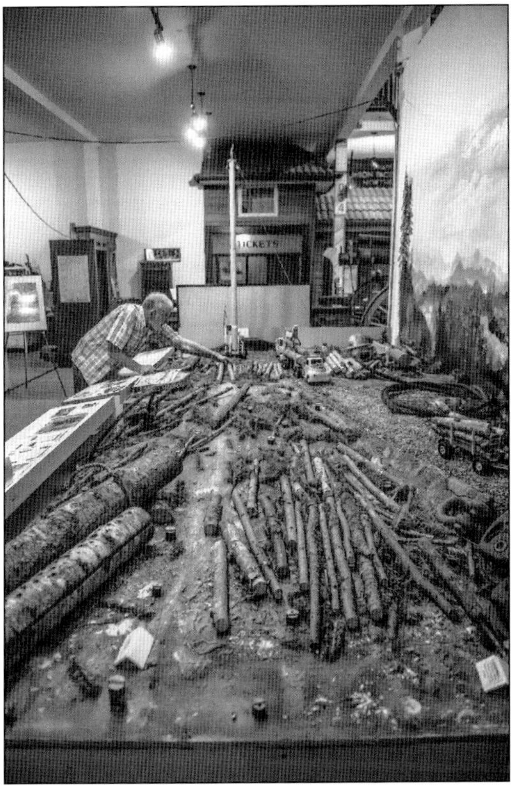

Byron Eager has spent hundreds of volunteer hours at the Aberdeen Museum of History constructing a miniature logging show as an educational display and a salute to the glory days of logging. On the right side of the photograph, director Dann Sears's mural shows a scene typical of what the loggers experienced daily on the job. (Brian Woodwick.)

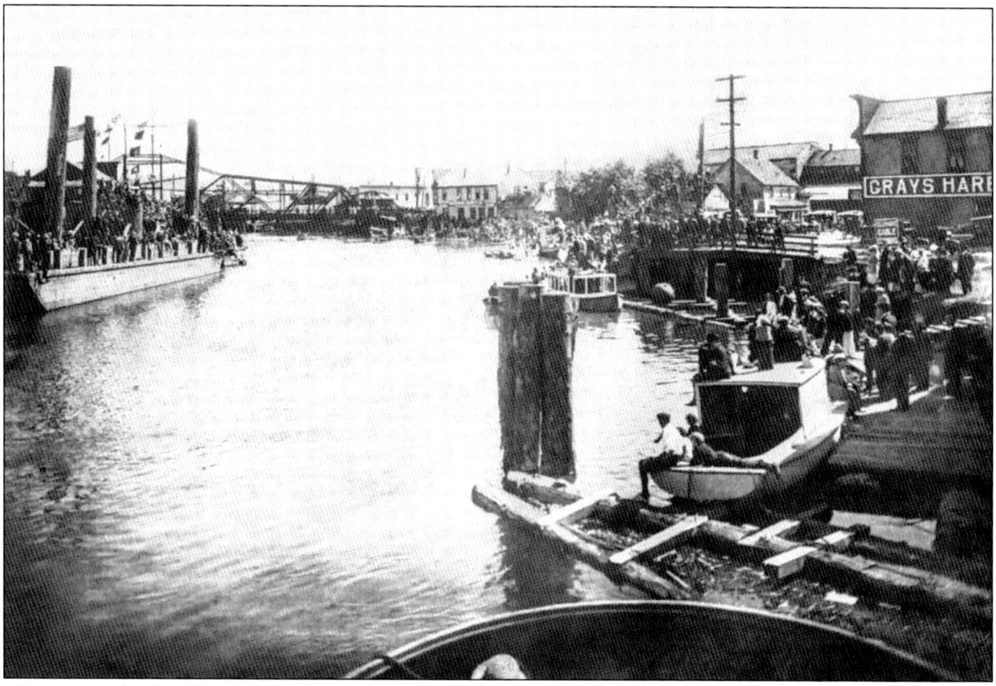

The annual Aberdeen Splash celebration on July 4 commemorates the early days of logging, when the area streams were full of dammed logs waiting for a late spring freshet to flood them downstream to tidewater and the mills. (AMH.)

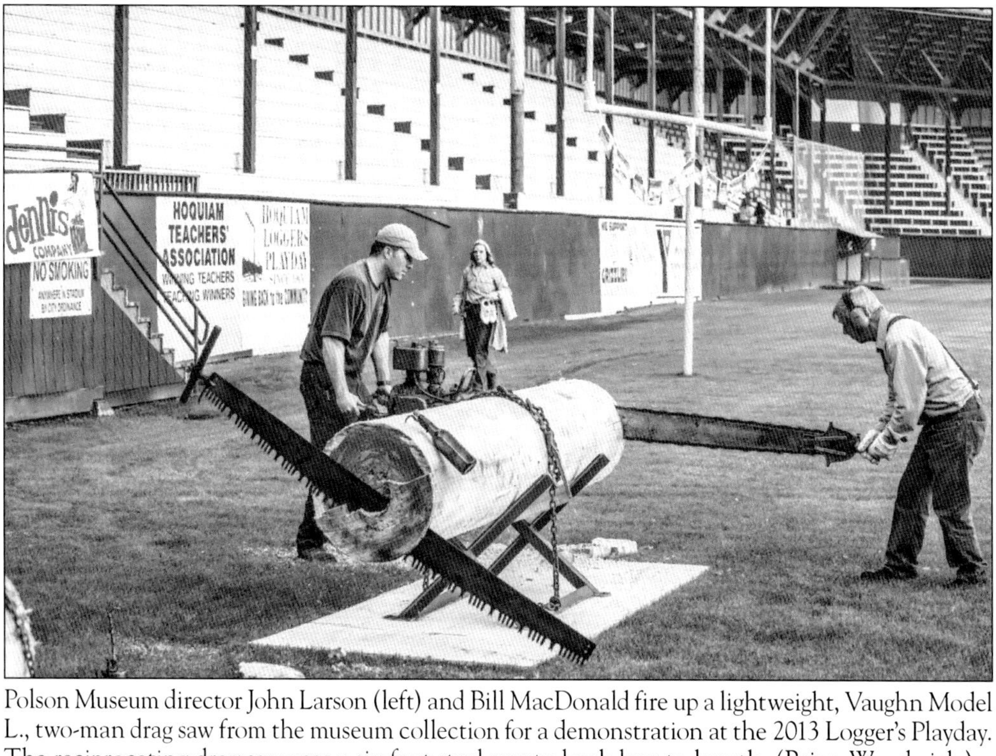

Polson Museum director John Larson (left) and Bill MacDonald fire up a lightweight, Vaughn Model L., two-man drag saw from the museum collection for a demonstration at the 2013 Logger's Playday. The reciprocating drag saw uses a six-foot steel saw to buck logs to length. (Brian Woodwick.)

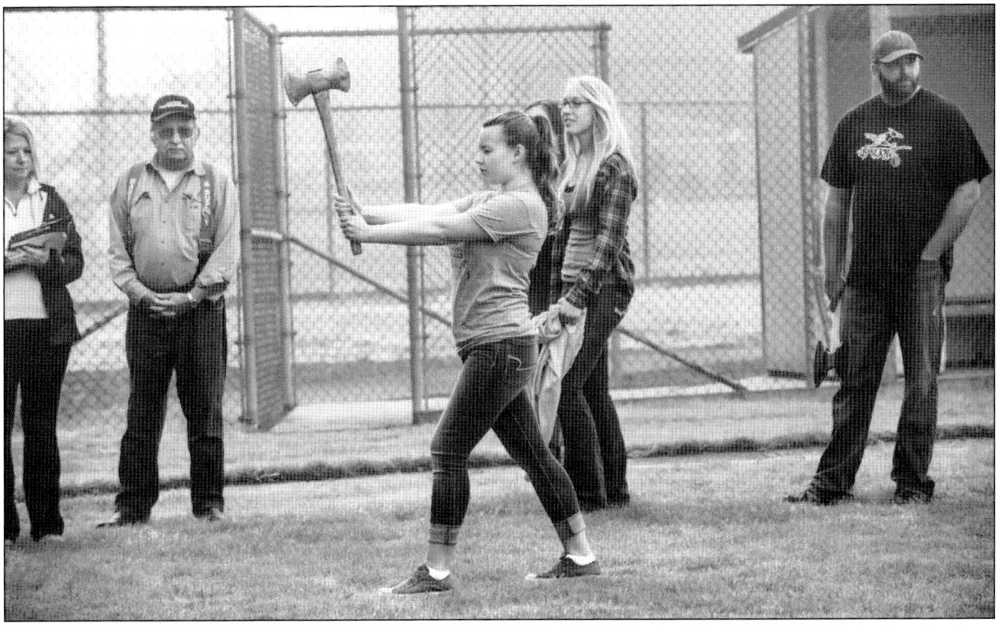

The ax-throwing contest is one of the events open to both men and women. This woman displays the concentration and muscle it took to win at the 2013 Hoquiam Loggers Playday. The 36-inch-wide target with five rings four inches wide establishes the scoring. Each player, standing 20 feet from the target, is allowed five throws. (Brian Woodwick.)

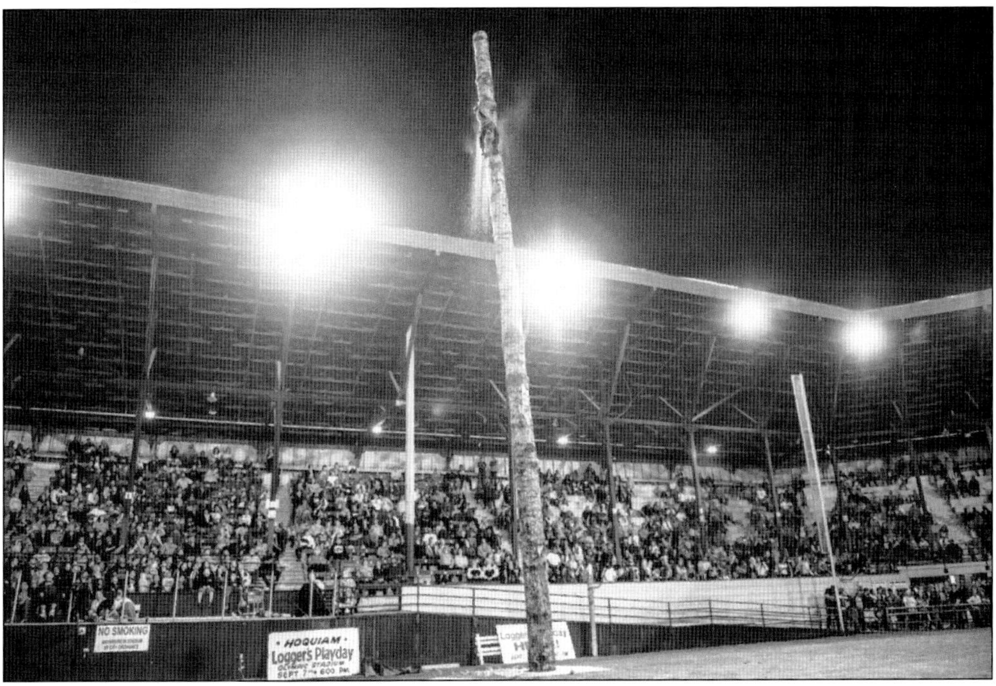

Thousands of spectators jammed the Olympic Stadium in Hoquiam to cheer on contestants in the 49th annual event, which displays the skills required daily by loggers in the woods of Grays Harbor. Local loggers compete with contestant loggers from other areas to attain the coveted top slot in the all-around event. (Brian Woodwick.)

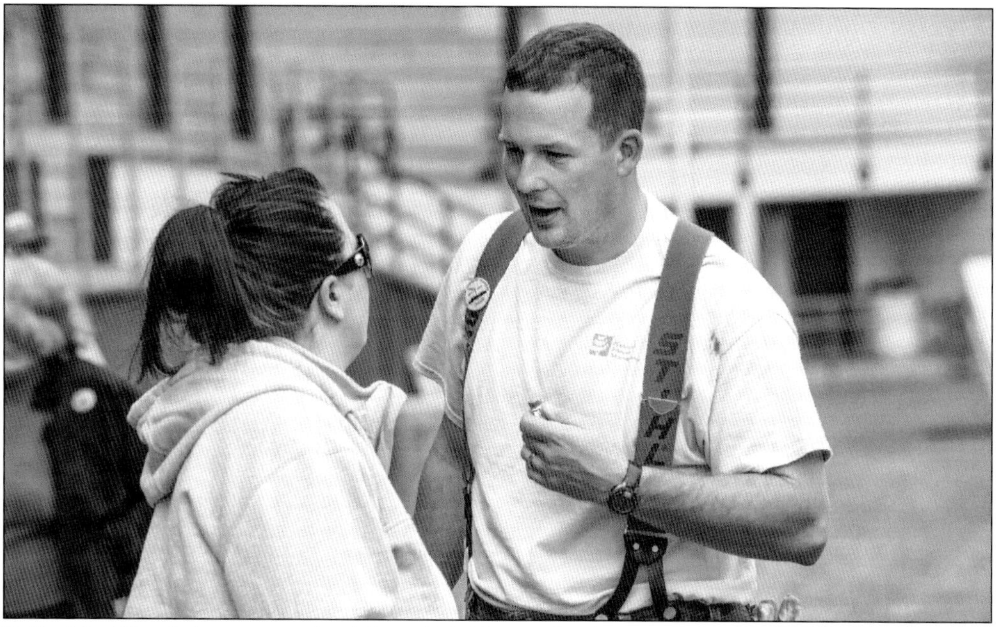

Bull of the Woods (superintendent) Jarrod Todd provides direction to a timekeeper at the Loggers Playday. He has been involved around the clock for years as a volunteer for the organization, which provides scholarships to students, numerous renovations for the civic stadium, and funds Loggers Playday expenses. (Brian Woodwick.)

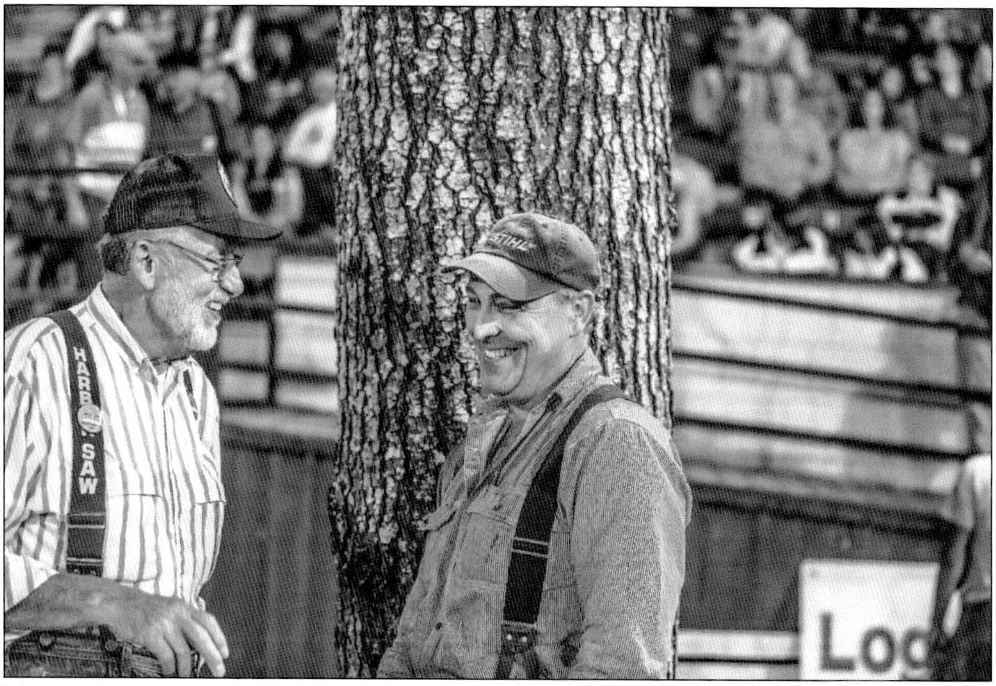

Master of ceremonies Paul Mackenzie (left) gives high climber John Rodius a hard time over their bet. Rodius wagered that he could smash a line of watermelons on the ground with the falling cut from topping the tree. Mackenzie disagreed. The loser donated $100 to the scholarship fund. (Brian Woodwick.)

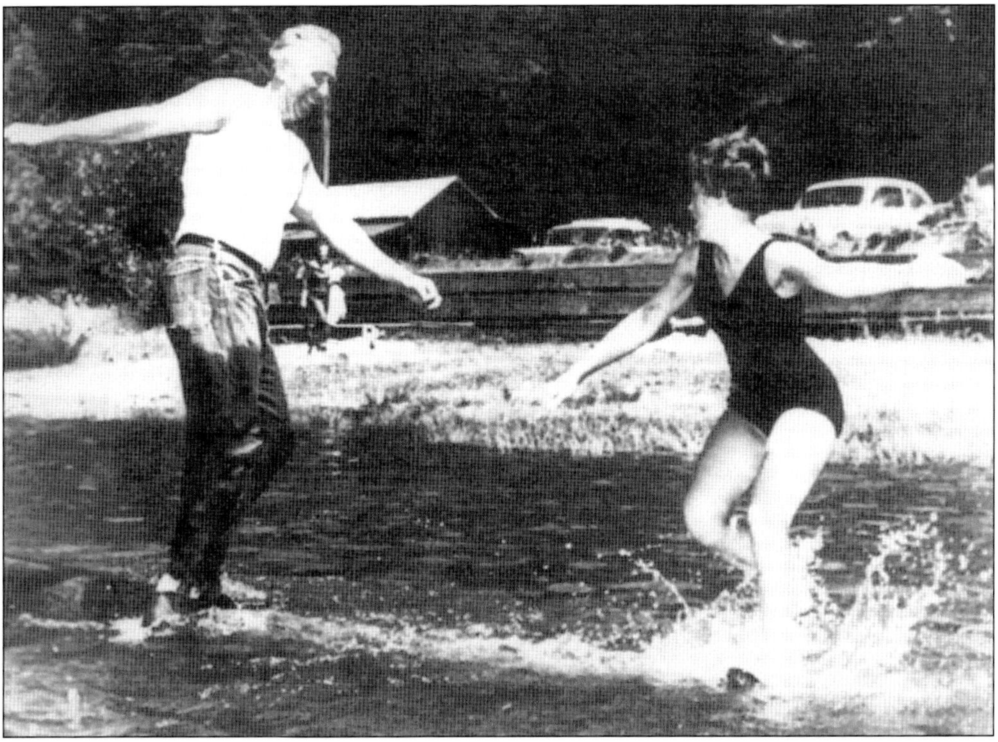

Russ and Diane Ellison, of Wishkah, were multiple world champion logrollers. Here, they perform at Lake Sylvia in Montesano in 1960. In 1961, they won the Trick & Fancy Division in the World Trials; Russ won the Old Timers (over 45) title, and Diane took home the Women's World Championship trophy for logrolling. (CHSGWC.)

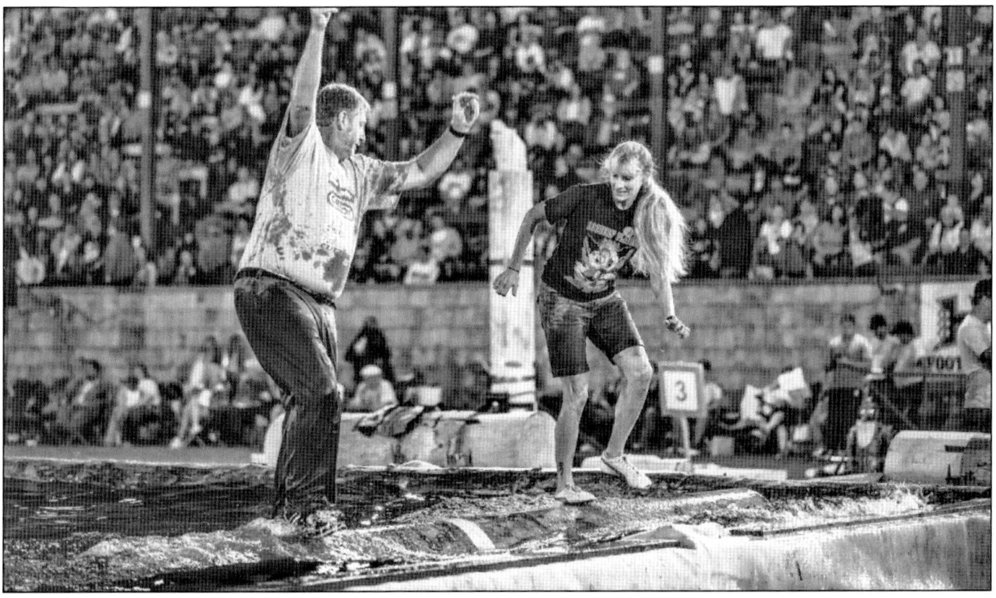

Exhibition logrollers Jim Dodge and Jenny Owens work to dump the other into the water during the 2013 Playday show time. While Midwestern loggers call it burling, Northwesterners call it logrolling. The ability to walk over and roll on logs was essential. (Brian Woodwick.)

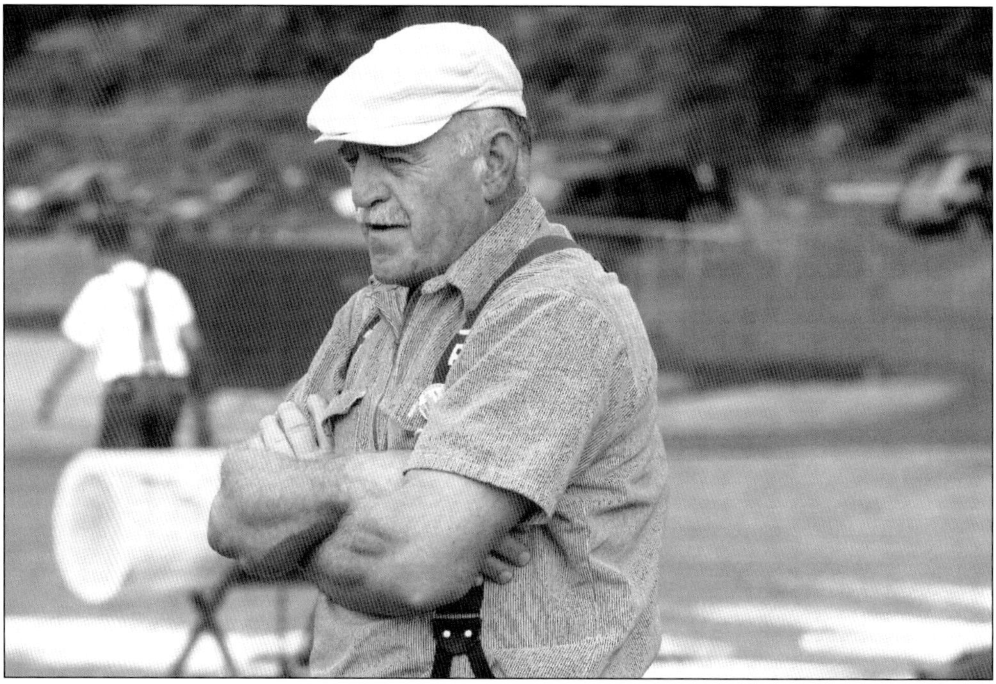

Lee Pickett studies the competition before winning the ax-throwing contest. Pickett and his brother Darrel are favorite longtime contestants who honed their logging skills in the North County mills and logging outfits. They come from a long line of timber workers. The brothers enjoy mentoring younger contestants. (Brian Woodwick.)

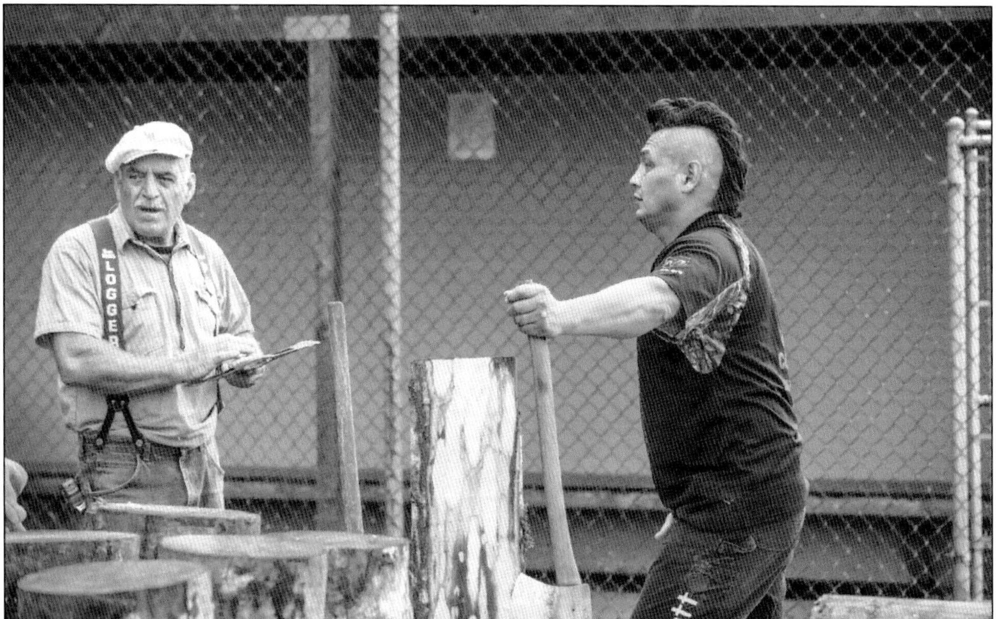

Lee Pickett (left) and David Moses Jr. discuss ax-sharpening techniques as they prepare to best each other in the ax-throwing event at Playday. The array of specially made axes used in the various events are razor sharp and are carefully maintained by their owners. (Brian Woodwick.)

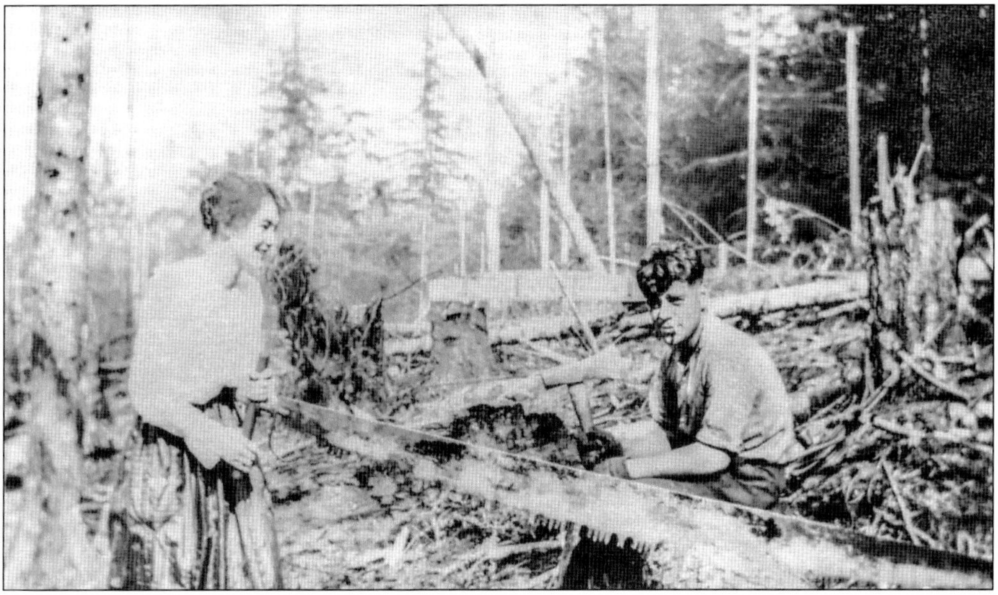

This young couple took time to pose with one of their homestead chores, little knowing that someday their log bucking would be an event in all Northwest logging contests. Many Harbor parents and grandparents do not have fond memories of hand-bucking the family's firewood. (CHSGWC.)

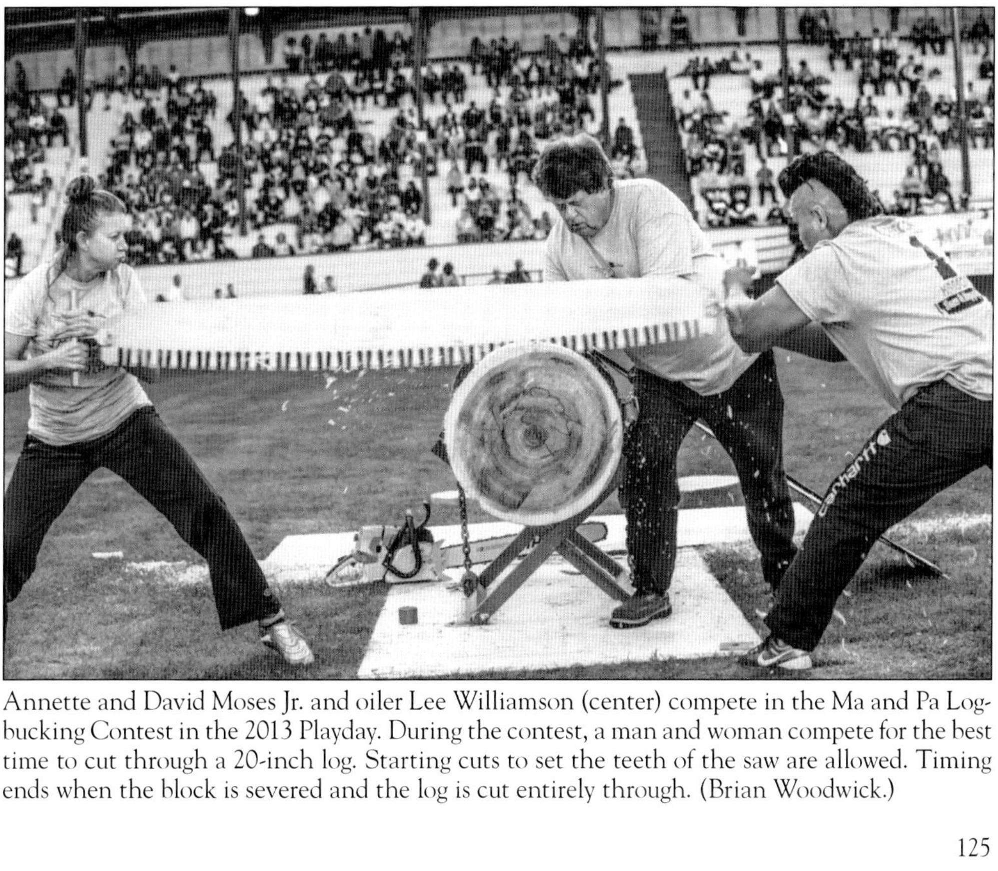

Annette and David Moses Jr. and oiler Lee Williamson (center) compete in the Ma and Pa Log-bucking Contest in the 2013 Playday. During the contest, a man and woman compete for the best time to cut through a 20-inch log. Starting cuts to set the teeth of the saw are allowed. Timing ends when the block is severed and the log is cut entirely through. (Brian Woodwick.)

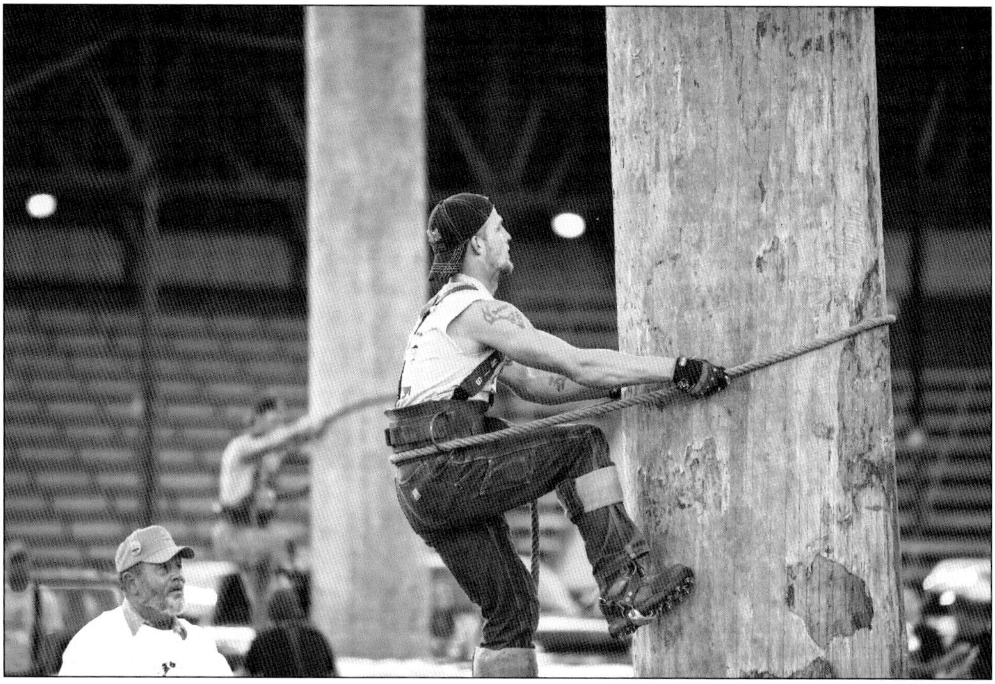

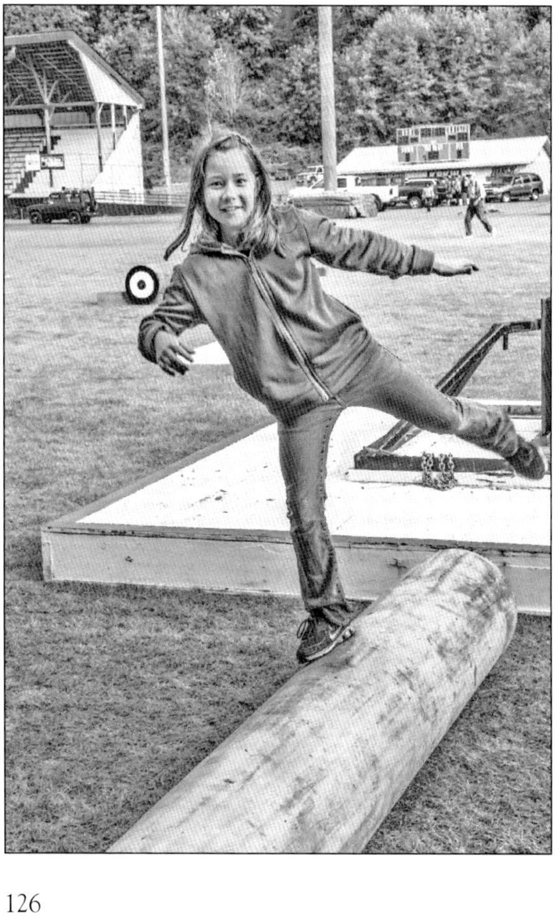

James McManus takes his mark prior to the high-climbing competition. Contestants scale a 90-foot-tall cedar spar pole and return to the ground while being timed. At the starting signal, contestants must have one foot on the ground and the other foot below the orange line marked on the sparring pole. (Brian Woodwick.)

This vibrant young girl knows exactly what she wants to compete in when she grows up—logrolling. She is pretty sure she has all the right moves to keep a log rolling and to knock the other contestant into the water, smiling all the time. (Brian Woodwick.)

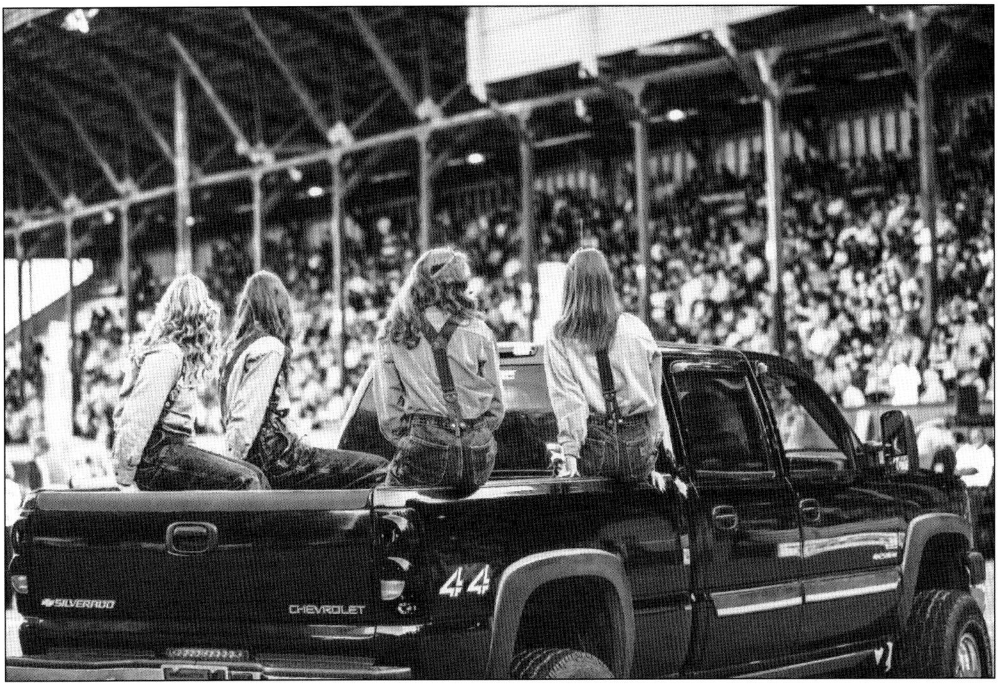

Many Grays Harbor teenage girls look forward to becoming queen and princesses of the Playday, riding in the parade and participating in other events. Here, the 2013 Loggers Playday Court faces the cheering crowd. Those living their dream were, from left to right, Brittany Kathryn Nations, Allie Drake, Mikaila Jackson, and Alexis Franklin. (Brian Woodwick.)

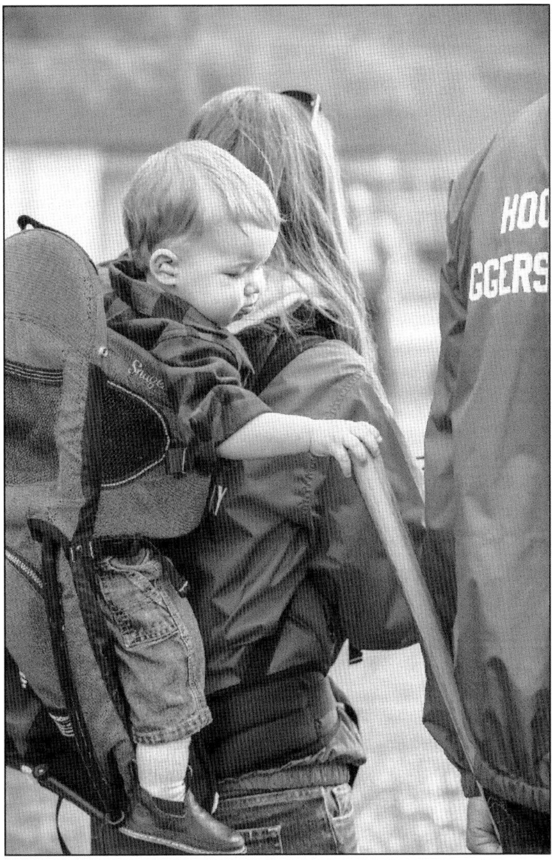

As this tyke shows, one just cannot be too young to want to be a logger. He grasps a contestant's ax handle as he is all togged out in a plaid shirt, red suspenders, jeans, and Romeo shoes. His dad was a contestant in the logging events. (Brian Woodwick.)

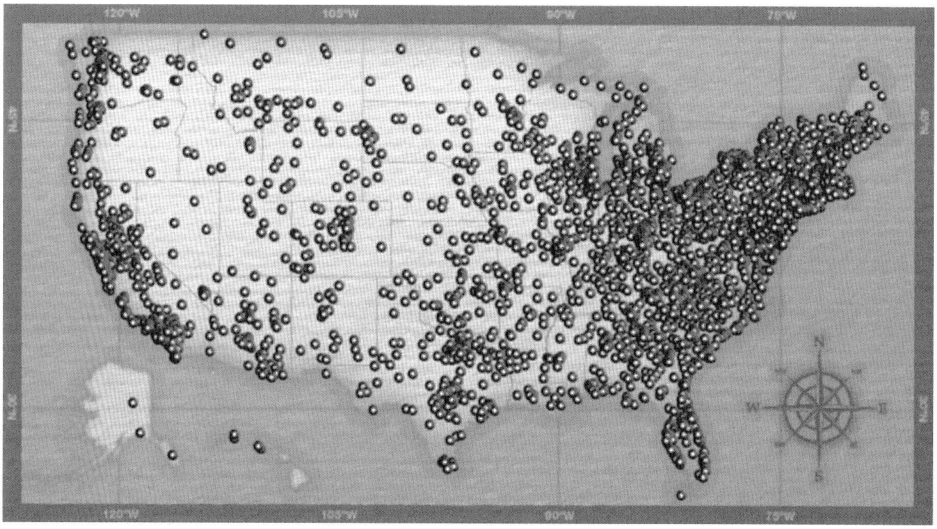